RESEARCH CENTRE

REPORT OF THE
ONE HUNDRED AND TWENTY SEVENTH ROUND TABLE
ON TRANSPORT ECONOMICS

held in Paris on 4th-5th December 2003
on the following topic:

TIME AND TRANSPORT

EUROPEAN CONFERENCE OF MINISTERS OF TRANSPORT

EUROPEAN CONFERENCE OF MINISTERS OF TRANSPORT (ECMT)

The European Conference of Ministers of Transport (ECMT) is an inter-governmental organisation established by a Protocol signed in Brussels on 17 October 1953. It comprises the Ministers of Transport of 43 full Member countries: Albania, Armenia, Austria, Azerbaijan, Belarus, Belgium, Bosnia-Herzegovina, Bulgaria, Croatia, the Czech Republic, Denmark, Estonia, Finland, France, FRY Macedonia, Georgia, Germany, Greece, Hungary, Iceland, Ireland, Italy, Latvia, Liechtenstein, Lithuania, Luxembourg, Malta, Moldova, Netherlands, Norway, Poland, Portugal, Romania, Russia, Serbia and Montenegro, Slovakia, Slovenia, Spain, Sweden, Switzerland, Turkey, Ukraine and the United Kingdom. There are seven Associate member countries (Australia, Canada, Japan, Korea, Mexico, New Zealand and the United States) and one Observer country (Morocco).

The ECMT is a forum in which Ministers responsible for transport, and more specifically the inland transport sector, can co-operate on policy. Within this forum, Ministers can openly discuss current problems and agree upon joint approaches aimed at improving the utilization and at ensuring the rational development of European transport systems of international importance.

At present, ECMT has a dual role. On one hand it helps to create an integrated transport system throughout the enlarged Europe that is economically efficient and meets environmental and safety standards. In order to achieve this, it is important for ECMT to help build a bridge between the European Union and the rest of the European continent at a political level.

On the other hand, ECMT's mission is also to develop reflections on long-term trends in the transport sector and to study the implications for the sector of increased globalisation. The activities in this regard have recently been reinforced by the setting up of a New Joint OECD/ECMT Transport Research Centre.

*

* *

Publié en français sous le titre :
LE TEMPS ET LES TRANSPORTS

Further information about the ECMT is available on Internet at the following address:
www.cemt.org

© ECMT 2005 – ECMT Publications are distributed by: OECD Publications Service,
2, rue André Pascal, 75775 PARIS CEDEX 16, France

TABLE OF CONTENTS

THE IMPORTANCE OF THE COST AND TIME OF TRANSPORT
FOR INTERNATIONAL TRADE

Alan V. DEARDORFF
The University of Michigan
Michigan
USA

Round Table 127: Time and Transport – ISBN 92-821-2330-8 - © ECMT, 2005

THE IMPORTANCE OF THE COST AND TIME OF TRANSPORT
FOR INTERNATIONAL TRADE

SUMMARY

Michigan, October 2003

* The author benefited from discussions on issues in this paper with David Hummels and
Robert Stern. The paper draws in part on Deardorff (2003b).

1. INTRODUCTION

Rather oddly, economists like myself who specialize in the theory of international trade have tended to ignore its cost. That is, like physicists assuming away friction in mechanics, we often assume that the transportation of goods from one country to another can be accomplished "costlessly" and, if time enters our models at all, instantaneously. We do not believe that this assumption is accurate in reality, but we hope that it provides, like many of the other assumptions that we make, a good approximation to the world. For it permits us to cut through cumbersome complexities to gain what we hope is greater insight into more fundamental economic relationships.

Recently, however, some have begun to see the dangers of this approach. Too much of what we see in the world and in our data cannot be understood if trade is assumed to be frictionless. Therefore, we have begun to examine more carefully what role such costs of trade may play in determining both the amount and the nature of the trade that takes place. One of the findings has been that simply accounting for the explicit costs of transportation is not enough to explain certain things, such as the rather low volume of trade that exists in the world. One possible explanation for this, in turn, is that the resource costs of transporting goods do not tell the whole story; that the time required to accomplish international trade is important as well, at least in some industries. Therefore, recent attention has been directed not only at the amounts that exporters and importers must pay for transportation of their goods, but also at the time they must wait for delivery.

In this paper, first, Chapter 2 reviews some of the evidence indicating that costs of trade may be too large to ignore. In Chapter 3, the various forms that trade costs may take are discussed, including both the resource and the time costs of transportation as well as other costs that may arise. Then in Chapter 4, some of the main implications of trade costs are examined, implications that are likely to be relevant for costs of all these sorts. Among these costs, the time costs may be the least familiar, and these are discussed further in Chapter 5, where it is noted that while the time required for transporting goods has become smaller, the importance of even small delays in many industries has become larger. These two facts interact to provide additional implications for patterns of trade. Most of this discussion is equally relevant, in principle, to all countries. However, there are some differences that exist among countries based on level of development, and these are examined in Chapter 6, before concluding in Chapter 7.

2. WHY TRADE COSTS MAY BE IMPORTANT

The frictionless world is well understood in theory, since it has occupied the attention of trade economists throughout most of the history of trade theory. If costs of trade are zero, then any good should fetch the same price everywhere in the world, and both producers and consumers everywhere will therefore face the same constellation of prices to guide their decisions. Immediately, we see some problems in matching this to the real world, evident to anyone who has paid attention to prices as they have travelled the world. The more obvious of these international price differences are for non-traded goods, and therefore are not evidence of trade costs[1]. Other price differences may reflect differences in taxation or regulation. But the main issue, since trade costs are obviously not really zero, is whether such price differences as do exist are large enough to matter significantly in their implications for economic behaviour.

One of the most famous implications of frictionless price equalization through trade is the Factor Price Equalization (FPE) Theorem, which Samuelson (1948) demonstrated in the context of the Heckscher-Ohlin, or factor-proportions, model of international trade. According to the Theorem, if several additional assumptions are satisfied -- perfect competition, identical constant-returns-to-scale technologies and sufficient international similarity of factor endowments that countries produce enough goods in common -- then international equalization of goods prices causes international equalization of factor prices. That is, if trade is frictionless, then it should lead wages, rents and returns to capital and education to be the same in all countries.

Again, FPE seems obviously not to hold in the real world, where, for example, wages differ across countries by factors of ten or more. However, this could be due to failure of assumptions other than frictionless trade. For example, Trefler (1993) showed that such wage differences could arise if technologies were to differ across countries in a simple way, otherwise preserving the implication of FPE for "effective" units of labour.

However, Trefler (1995) found shortly thereafter that, even with such differences in technologies, frictionless trade should find countries trading far more than they actually do. For this he examined another implication of the Heckscher-Ohlin Model, called the Heckscher-Ohlin-Vanek Theorem, after Vanek (1968). The HOV Theorem says that the "factor content" of a country's trade -- the amounts of factors embodied in its net exports -- should be equal to the difference between its factor endowment and the average factor endowment of the world. In fact, actual factor content of trade amounts to only a tiny, even negligible fraction of what that theory predicts. Trefler (1995) called this the Mystery of the Missing Trade. This is a mystery which would easily be solved if trade itself were also small, perhaps because of high costs of trade. And if that were the explanation, it would certainly confirm that trade costs were high enough to matter, since they would have invalidated one of the central results of trade theory.

A final body of evidence also indicates that trade costs may be quantitatively important. Perhaps the most empirically robust relationship in international economics is the Gravity Equation, an equation that relates the volume of trade between pairs of countries positively to their economies' sizes and inversely to the distance between them. Gravity equations have been estimated on trade flow data ever since Tinbergen (1962) and Pöyhönen (1963), and they have fitted the data remarkably well. Estimates of gravity equations consistently find an elasticity of trade with respect to distance of around 1.0 or a bit less, which means that, other things being equal, if two countries were twice as far apart they would trade approximately half as much. In a world without any frictions at all, there would be no reason for distance to matter at all, and so the fact that it does matter quite strongly suggests that trade costs are positive and rise with distance, presumably at least in part due to the costs of transportation.

The gravity equation has been used increasingly in recent years as a benchmark for trade flows, the point often being to explain departures from the simple gravity equation in terms of other variables that might be thought to matter for trade. One such exercise was done by McCallum (1995), who applied the gravity equation to trade among provinces and states of Canada and the United States. What he found was that, not only does distance matter, but so does the national border between the

Round Table 127: Time and Transport – ISBN 92-821-2330-8 - © ECMT, 2005

two countries. That is, there is less trade between an arbitrary province of Canada and a state in the United States than between that same province and another Canadian province of the same size and distance away as the state. This suggests that there are trade costs associated not only with distance but also with the need to cross a border. Similar evidence has been found elsewhere in the world, by Helliwell (1998) and others.

3. TYPES OF TRADE COSTS

The most obvious and probably the most important cost of trade for most products is the resource cost of transportation: what a trader must pay for carriage of its goods from one place to another by some particular mode of transportation. This cost, of course, varies with the distance travelled, with the weight and bulk of the item carried, and perhaps with the care that must be taken along the way to avoid damage or spoilage. There is also the related cost of loading and unloading the good at both ends of the trip, and perhaps some extra costs of getting the good to and from the shipping terminal. All of this can be distinctly nonlinear, including fixed costs which are unrelated to distance and/or to the size of the shipment, as well as costs per unit distance or size which may also vary due to economies of scale or due to discontinuities related to container size or geography.

All of these complications are undoubtedly important for the firms engaged in the actual trade, but they may not matter much for the broader questions considered here, of the effects of these costs on overall trade flows and the allocation of production. Those in trade theory who have attempted to model transport costs have usually made them very simple. The classic example, often imitated, was Samuelson (1952), who assumed "iceberg" transport costs: a fraction of the transported good itself is used up in transit, and that is the only cost. This assumption has the advantage of capturing in a very simple way a transport cost that depends on the amount transported and which can be made proportional to distance. And it has the further advantage that, because the only resource used is the traded good itself, its value relative to that good does not vary with the prices of other goods and factors. Nor does it give rise to a demand for anything in another market, which is an advantage in the general equilibrium type of model that is necessary to examine the main issues of trade theory.

There are other costs of trade besides transport which are realistically of some importance, although their implications seem likely to be similar to those of transport costs, and it may therefore not be necessary to consider them separately. One such is insurance which, like transport, is likely to vary positively with both the size of the shipment and the distance shipped. Unlike transport of most products, though, insurance cost will also increase with the value of the shipment, independently of its weight or bulk. Still, the cost of insurance is so similar in its determinants and effects to the cost of transport that the two are often lumped together.

Another trade cost is the cost of trade financing. This would exist even for transactions within a country, as long as there is a delay between the incurring of production costs and the receipt of payment for the final sale. Since trade adds to this delay, it also adds to the cost of financing, in a way that depends on time, as we will discuss below. However, international trade also involves other financial complications that do not arise for domestic trade. To the extent that financial markets are not well integrated across national borders, transactions may require dealing with different banks and therefore require associated fees. And, most importantly, trade often requires conversion from one national currency to another, and there are both fees and uncertainty associated with that. Normally, these financial costs should depend almost solely on the value of a shipment, not on other dimensions of its size and not on the distance over which trade takes place. However, since nearby countries are likely to have more integrated financial markets and reciprocal arrangements between their financial institutions than countries which are far apart, these costs too may be loosely related to distance.

There are undoubtedly other costs of trade that the author is not thinking of, in addition to those associated with time which will be discussed below. The author's confidence in his own ignorance was heightened by a paper he heard presented some time back which drew attention to the costs of operating within two different legal systems, Turrini and van Ypersele (2002). The point was that any international transaction straddles borders, and each nation has its own legal system to be used if the transaction goes awry. Those who trade must be prepared for misadventure and prepared perhaps to use the courts of another country to pursue their claims. But that in turn requires legal expertise which their domestic operations can do without, thus incurring additional expense. Like the cost of financing, these legal costs seem related mostly to the value of the transaction, but they are also likely to rise with the actual or cultural distance between countries. Now this is an aspect of trade costs that the author would never have thought of himself, and it suggests to him that there may be many other such costs as well.

4. TIME COSTS

Turning now to the role of time, it enters the story most obviously because it takes time to get a good from one country to another. A large part of that time may be used in transport, but additional time may be required to load and unload and to process a shipment through customs and overcome any regulatory hurdles. The question, for all of these sources of delay between production of a good and its delivery to the customer, is how costly does the delay turn out to be?

The cost of time will be minimized, though it will not be zero, in a world where nothing changes over time. Our simplest economic models are static ones, in which time does not enter at all. These may be thought to correspond to an imagined reality in which flows of supply and demand occur per unit time and do not change over time, or else are correctly anticipated. In such a world, if in fact there is a delay between the moment a good leaves the factory gate and when it reaches the demander, the delay can be accommodated by simply producing and shipping ahead of time by the amount of the delay. In that case, the only cost in such a static world is interest cost. That is, the supplier incurs the cost of production before the good is shipped, but is paid for the shipment only at delivery[2]. The cost of this time delay is the interest paid or foregone by the shipper in order to pay its workers and suppliers before being paid by the customer.

However, this interest cost is surely rather small. Suppose that real interest rates are, say, six per cent per annum, and that it takes two months to get a product to a foreign demander after it leaves the factory. Then the interest cost of this delay -- one-sixth of a year times six per cent -- is just one per cent of the value of the shipment. This is not necessarily negligible, but it is small compared to many other costs of trade, such as tariffs and transport costs in most industries. And it is very small, as we shall see below, compared to the expense that many traders are willing to incur to reduce such delays.

Anderson and van Wincoop (2003), for example, in their excellent survey of the literature on trade costs, put the average cost of transportation at 21 per cent. Therefore, while these interest costs are undoubtedly part of the story of trade costs, I suspect that they are only a small part.

Costs rise when the sizes of supply or demand become uncertain and are therefore impossible to anticipate precisely. In such cases, staying with the example begun above, the supplier does not know two months in advance how much to produce to meet the later demand. In the absence of inventories, if demand turns out to be higher than previous production, then either sales are lost or demanders must be persuaded to wait for delivery, both of which options are likely to be costly. If demand turns out to be lower than production, then the good accumulates as inventory, which may not be so bad as long as demand does not disappear entirely. Therefore, because this penalty for producing either more or less than demanded is asymmetric, firms routinely deal with such uncertainty by holding inventories.

The cost of holding inventories includes several parts: interest, storage and depreciation, all of which vary greatly across products. The interest cost of holding inventories of some products may be even smaller than the interest cost of the time lost in shipping. Suppose, for example, that uncertainty in demand is such that, as of the moment two months prior to delivery when the decision to ship to a market must be made, the uncertainty about demand amounts to twenty per cent either way. That is, whatever may be the expected value of demand, the trader knows that actual quantity demanded may turn out to be twenty per cent above or below this. To avoid coming up short, he must either ship, or hold (if inventories have been held over from the past), an extra twenty per cent above the expected demand. Holding this amount for two months incurs an interest cost, for the same numbers as above, equal to one-fifth of one-sixth of six per cent of the expected value of demand. Only in the event of what the author would call one hundred per cent uncertainty -- the possibility that realized demand may be twice its expected value -- will this interest cost of inventories be as large as the interest cost of shipping.

Therefore, whether the interest cost of holding inventories is large or small depends on the size of inventories and thus on this uncertainty of demand. This in turn may depend on the homogeneity of the good as produced by a particular firm. If it produces only a few distinct product lines, then demand for each may be reasonably predictable. But if it produces a wide variety of variations of its product, even if those variations are slight, it will need to hold inventories of each variety, and uncertainty of demand for each will be larger, perhaps much larger. If a particular variety, for example, sells only intermittently, then the uncertainty relative to its expected sales during the period required to replenish inventories, could be several hundred per cent. Thus a multi-product firm may need to hold inventories of all its varieties close to demanders. The cost of this will be substantial and will rise with the time required to replace those inventories, after a sale, and thus with the time that it takes to trade.

Storage and depreciation seem likely to vary even more across products. Take storage. Some goods take up little space and require little special care, so that storage costs are minimal. Others are bulky, heavy, sensitive to temperature or humidity, or simply valuable enough to be tempting to thieves. These therefore require costly space or special structures and protection to accommodate them. These costs, for some products, may be much larger than the interest costs discussed above. On the other hand, they may still be small relative to the resource costs of transportation. Because all of the reasons why a good is expensive to store are likely also to make it expensive to ship, since shipping is, in effect, storage combined with the added expense of movement.

Depreciation, as that term is normally understood, seems likely to be even smaller. If it merely means the loss of usefulness and value as a good wears out through use, in fact, it may be zero during the time taken to trade it, since it is not used. And even if perceived depreciation depends on the calendar time that has elapsed since the good was produced, this too is likely to be small. A good that depreciates to zero value over ten years, for example, would lose only 1/60th of its value over two months, if that were the time needed for trade.

However, the author uses "depreciation" to include any reason for loss in value over time, not just wearing out through use. An increasing number of traded products are perishable, for example, including fruits and vegetables and cut flowers. Many of these lose their value in a matter of days, some even hours, a speed which rendered them simply non-tradable before innovations in transport technology enabled them to be shipped great distances at comparable speed.

Also under the heading of depreciation, the author would include the sometimes rapid obsolescence which occurs as changes in technology or fashion replace models or styles with new ones, and which thus forces down the market value of old designs well below what seems to be their value in use. Consumer electronic products, for example, appear with new features every few months, and when they do, the previous models which lack these features, even though they are still capable of performing as well as ever, drop precipitously in market value. Or, in another important example examined recently by Evans and Harrigan (2003), the clothing industry is subject to frequent changes in fashion which similarly turn last month's hot selling design into this month's candidate for the sale bin. Evidently such clothing provides a service to its users far beyond protection from the elements, since its ability to keep purchasers warm and dry is unaltered by changes in fashion, yet the market value drops to near zero.

What is crucial about these changes, for the purpose here, is not just that value declines with time, but that it is impossible to anticipate that decline by producing and shipping early. In contrast, demand for some products is seasonal, and demand disappears as the season passes. But if the time to market increases in such a case, and if demand is predictable, then a firm need merely produce that much earlier in order to meet the demand on time. Interest and storage costs are increased, but depreciation cost is not. In contrast, what is important for this sort of depreciation is that the nature of what will be demanded cannot be known long enough in advance in order to produce the desired product this early. If fashion decides only in November what the hot colour will be in December, then a firm that needs two months to produce and ship the desired colour will be out of luck. And whatever colour they do produce in October may lose value a month later if that colour too goes out of style.

To sum up, trade costs fall into the following categories:

- Costs which do not depend on time to market
  - Resource cost of transportation
  - Insurance
  - Financial costs of exchange
  - Other (legal costs, etc.)
- Costs which do depend on time to market
  - Interest
  - Storage
  - Depreciation
    - Spoilage
    - Obsolescence.

Although the word "transportation" appears in only one of these categories, it is critically relevant in much more. For all of the reasons that time to market matters for cost, the main way of reducing that cost is to choose a means of transport which is faster.

Indeed, it is this choice between slower and faster means of transport which has given us our best indication of the importance of time. David Hummels (2001) has used the costs of various modes of transportation to infer the costs of time from the amount that firms are willing to pay to reduce it, most obviously by choosing expensive air transport over surface modes. He found, on the basis of this evidence, that a one-day delay in shipping imposes an average cost equivalent to a 0.8 per cent tariff.

In what follows, we will now address the ways in which these costs of trade are likely to matter, not just for those engaged in the trade themselves, but for their economies more broadly. In this discussion, time costs are both typical and special, in that some of their effects are the same as those of other sources of cost of trade, while others of their effects are peculiar to the nature of time. The discussion is therefore split into two sections, addressing the generic costs of trade first and then turning separately to time.

5. EFFECTS OF TRADE COSTS

Costs of trade may be avoided by not trading. Therefore the largest effect of trade costs is that they reduce the volume of trade. How costly that is, in turn, to the welfare of countries depends on how much those countries depend on trade. A country in a remote location, far from world markets, or one which is economically remote because natural barriers or lack of infrastructure make it difficult to reach those markets, foregoes what would otherwise have been its gains from trade. Whether those gains are large or small depends on the resources already available in the country's location and its ability to use those resources to sustain and enrich life. The economic hardship imposed by high trade costs is largest in places where resources are not well suited to providing the needs of a population, even though they may provide something that would be valuable on world markets. Such places are naturally empty of population when trade costs are high, but they can become thriving metropolises as trade costs fall.

We are accustomed, in trade theory, to arguing that artificial trade costs such as tariffs do their damage by distorting markets. That is, a tariff on an imported good raises the price on domestic markets and both discourages demanders from purchases they would otherwise have made and encourages domestic import-competing producers to produce quantities which would otherwise have cost more than they were worth. Both of these distortions of behaviour are costly, as economic theory has spelled out carefully, and the result is that a tariff imposes a "dead weight loss" on the country that uses it.

Trade costs of the various sorts addressed here have the same effects on demanders and suppliers as tariffs. Here, though, it would be wrong to call these "distortions" of behaviour, since the responses are appropriate to costs which are real, not just from the perspectives of the demanders and suppliers themselves, but for the economies in which they live. But the fact that they are not distortions does not make them any less costly. On the contrary, a tariff is costly to society only to the extent that it does alter behaviour, and the bulk of the cost to the individual trader -- the tariff revenue -- is not a cost to society as a whole, but simply a transfer to another part of the economy, its government. In the case of real costs of trade -- in which the author includes both the resource costs and the time costs discussed above -- the costs accrue to nobody, or at least not as a net benefit[3]. Thus the cost to society of real trade costs is always larger, and often much larger, than the costs of a tariff of equal size[4].

This is especially true when the size of the barrier itself is not too large, so that a good deal of trade takes place in spite of it. In that case, the large amount that a government may collect in a tariff, which is not part of the net loss to society from the trade barrier, becomes a part of this deadweight loss when the barrier is, say, an equal sized transport cost. If, on the other hand, we compare a tariff and a transport cost which are both so large as to prohibit trade entirely, then their welfare effects are the same.

So, the most important effects of trade costs are simply that they reduce the amount of trade, forcing countries to rely more fully on their own resources and depriving them of the many sources of gains from trade. The gains from trade include especially the gains from specializing in accordance with comparative advantage and exchanging the fruits of that specialization for things that other countries can produce relatively more cheaply. But the gains also include the many benefits from access to larger markets, both for producers and consumers, who benefit from lower costs due to economies of scale, from greater variety of products to match their needs, and from lower prices due to competition among a larger pool of competing suppliers.

In addition to depriving a country of the overall gains from trade, trade costs also matter for the nature of what trade remains. That is, a country which might have had a comparative advantage in one good if trade were costless, might export instead a very different good if transport costs play an important role. There are at least two reasons for this, one purely geographical and the other depending on properties of technology.

The first reason appears in Deardorff (2003a), called "local comparative advantage." The point is that the nature of a country's comparative advantage depends on which other countries it competes with, and this in turn depends on the size of the costs of trade. If trade costs were close to zero worldwide, for example, a moderately developed country such as Mexico might be able to produce and export certain moderately capital-intensive goods to the world market. But if trade costs are high and depend importantly on distance, then the more important determinant of Mexico's trade will be her proximity to the US market, relative to which Mexico is not well developed at all. Therefore, trade costs may dictate that Mexico will continue primarily to export labour-intensive goods to its rich local neighbour. Similarly, Argentina, though labour-abundant on a global scale, may be capital-abundant relative to her South American neighbours and therefore export various capital-intensive goods for which transport costs are high.

This effect of trade costs on the pattern of trade is of intellectual interest to trade economists such as myself. But it can also be of considerable practical importance for participants in the world trading system. When trade costs fall with technological progress, patterns of comparative advantage will change too, and industries that may have prospered serving their local international neighbours may find that these industries are no longer viable. Such changes represent opportunities, as well, but those opportunities are often of little help to those employed in the industries that shrink.

More broadly, as suggested by the examples above of labour-intensive exports from Mexico to the US and capital-intensive exports from Argentina to its neighbours, trade costs and geography may in part determine which groups in a society gain and lose from trade. A standard result of the Heckscher-Ohlin model of international trade is the Stolper-Samuelson Theorem, which says that trade benefits a country's abundant factors and hurts its scarce factors. But with high trade costs and local comparative advantage, these concepts of abundance and scarcity need to be interpreted relative to the neighbouring economies with which trade is mostly likely to take place. Thus, whereas globally all of South America may be thought of as labour-abundant, so that trade liberalization ought to tilt the income distribution toward labour at the expense of capital, local comparative advantage may suggest that the countries of South America will be more diverse, with the vested interests in trade lying with capital in some countries and with labour in others.

There is at least one more way that trade costs can matter both for the volume and for the pattern of trade, and that is through the technology of intermediate inputs and the fragmentation of production. It has always been true that some intermediate inputs are traded. This is especially true of primary products and raw materials which are often available in only certain places in the world but need to be used elsewhere. In recent years, the role of intermediate inputs has increased, however, as it has become possible to split up, or fragment, production processes into stages that need not all be done in the same location. When fragmentation occurs, goods in process are passed along from one location to another for further processing, taking advantage of whatever other inputs, such as labour or specialized expertise, may be cheaper or more abundant in different locations.

All such input trade is possible only if the costs of trade, especially transport costs, are low enough to make it profitable. For traditional raw materials, transport costs are unlikely to prevent trade entirely, since the inputs are necessary and their cost, including costs of trade, will be incorporated into the price of the final product. But fragmented production processes and the resulting input trade compete with their integrated alternatives, and they therefore happen at all only if trade costs are sufficiently low. Indeed, most would agree that the increase in fragmentation and much of the resulting growth of trade in recent years has been the direct result of falling costs of trade. See Yi (2003) for a related account of how such "vertical specialization" has responded to falling trade costs in causing international trade to grow much more rapidly than GDP.

But input trade in the presence of trade costs has implications not just for the volume of trade but also, critically, for the locations of production and the geographical pattern of trade. If trade costs are high, little input trade takes place except for raw materials, and producers who need those raw materials tend to locate close to their source or, if not there, close to transport hubs such as ports where trade costs can be minimized. If trade costs were zero, on the other hand, location of production and trade would depend entirely on the costs of primary inputs such as labour, capital and land with, say, labour-intensive goods produced in labour-abundant countries and capital-intensive goods produced in capital-abundant countries, just as the Heckscher-Ohlin trade model predicts. But if trade costs are both positive and not too large, then producers will benefit from close proximity to other producers who either produce the inputs for them or use their product as inputs themselves. These are the "backward and forward linkages" which lie at the heart of the New Economic Geography of Krugman (1992) and others. These linkages give rise to agglomeration, the bunching of economic activity in cities and regions.

Agglomeration may lead to concentration of large amounts of activity within a single country, especially in a single industry, or in a group of related industries. In this case, the country will export the products of such industries, while other countries may be forced to specialize in sectors where traded inputs and the benefits of agglomeration are not so important. Alternatively, if trade costs are somewhat lower, a form of much looser agglomeration may occur across country borders, with groups of countries and whole continents attracting multiple interrelated industries while other continents are left behind.

Thus trade costs can be a critical factor in both allowing some countries to prosper and others to remain mired in poverty. Which countries fall into which category seems to depend in large part on accidents of history. Certainly, once a country is far removed from the bulk of economic activity taking place elsewhere, costs of trade can make it hard for it to overcome that barrier. And correspondingly, if trade costs can somehow be brought down, even further than they have already fallen, then this may reduce this disadvantage and permit poor countries whose economies are otherwise sound to make progress by taking greater advantage of international trade.

6. THE ROLE OF TIME

Time costs are just one of the many costs of international trade, and they contribute to all of the effects discussed above. Their contribution varies across industries, which differ in the importance of time for their markets. It is no accident that some of the earliest examples of international trade were in spices, which retained their value during the prolonged long-distance journeys required for transport over land. Likewise, relatively homogeneous primary products were later traded by ship, the traders confident that the markets for these products would last long enough for them to cross an ocean. Manufactured goods, on the other hand, were harder to trade over long times and distances except when they became sufficiently standardized for traders to be confident that demands for them would not disappear before the trade could be completed. As long as manufacturing meant artisans crafting products to the individual tastes of their customers, it was the time required for trade perhaps more than its resource cost which kept these goods non-traded.

In today's world, air transport permits many goods to be traded so much faster than a century ago that the landscape of international trade has radically altered. Faster still is the trade in a small number of services which can be transmitted electronically and, therefore, essentially instantly. It is these technological innovations which have turned some previously non-traded goods and services into traded commodities, such as cut flowers and banking services. They have also caused visible transformations of the economies of certain countries which have managed to take advantage of these changes.

At the same time, as already discussed above, the pace of change in both technology and consumer tastes has also picked up, so that remarkable innovations in speed are not always fast enough for the new markets. For many products, it is still worth incurring high production costs in order to be close enough to a market to respond quickly to its changing demands. This is certainly true in the early stages of a high-technology product, as was noted years ago by Vernon (1966) in his Product Cycle theory of trade. But it is also true today in the seemingly more mundane world of the apparel industry, where the importance of time has been documented in a fascinating study by Evans and Harrigan (2003).

The Evans and Harrigan story is instructive. Based on information from the apparel industry, they divide products into "replenishment" goods and "non-replenishment" goods. The former are those for which fashion changes so rapidly that firms cannot stock up for an entire selling season ahead of time without risking that products will lose their appeal and remain unsold. So they replenish during the season after they have seen which products sell and which do not. In this situation, they cannot wait the long periods that might be required for delivery from the least-cost production location and instead they produce closer to the US. Since production costs still matter a good deal, however, they do not produce *in* the US, but rather take advantage of low-cost (but now not lowest-cost) countries which are close by, such as Mexico and Latin America. Thus the Evans-Harrigan model

Round Table 127: Time and Transport – ISBN 92-821-2330-8 – © ECMT, 2005

predicts that the location of production of apparel products will vary systematically with distance from the US, the replenishment goods being produced close by and the non-replenishment goods produced further away. Their empirical analysis supports this prediction strongly.

This story has another implication, which could be obtained from the presence of trade costs more generally, but which is given particular force by the importance of time: Wages are likely to fall as distance from major consumer markets, such as the US, increases. The reason is that, in order for time-sensitive products to be produced at all at great distance from these markets, they must be attracted there by compensatingly lower wages. Of course, not all products are time-sensitive, even in the model of Evans and Harrigan, and it is possible that, beyond a certain distance from markets, wages will fall no further for greater distances, all such locations specializing in time-insensitive goods. But it will still be true that countries within that limit will be able to pay higher wages in return for the faster delivery that they offer.

This, of course, is not all that different from what one would find if time were not a factor and it were merely the resource cost of transportation and trade that determined location of production. The lesson, then, is not that the importance of time necessarily causes major changes in the patterns of trade and production that we see in the world. Rather it is that the importance of time tends to reinforce the patterns that we see already due to distance and resource costs of trade and that, as resource costs fall with improvements in transportation, time costs may replace them as the most important cost of trade for at least certain products.

On the other hand, the evidence in Hummels (1999) suggests that resource costs of trade have not in fact fallen all that much over the last half century, even as the volume of trade has grown substantially relative to production. Hummels (2001) goes on to suggest that falling time costs may have replaced falling resource costs as a stimulant to trade, as air transport has cut the times needed for trade in certain products to a small fraction of what they were before. Thus, the growth of trade may be accounted for less by falling resource costs of transportation than by its increasing speed.

The phenomenon of fragmentation, or outsourcing is mentioned above as a form of trade for which trade costs are especially important because of the additional trade which takes place between stages of a fragmented production process. Time can be especially important here, since delays in delivery from one stage to the next can shut down the whole chain[5]. Much has been made of the "just-in-time" production methods of the Japanese, introduced in the 1980s, with the usual interpretation that these methods lessened the costs of holding inventories. This they certainly did, but the more important contribution of these methods may have been the flexibility that they offered, especially as production became fragmented across locations, to respond quickly to changes in the need for inputs at various stages of production.

In short, it appears that the role of time in trade is becoming increasingly important, at the same time that the ability of technology to reduce the time required for trade has improved. The resource costs of trade will never be negligible, but the importance of time seems to loom ever larger in many products.

7. TIME AND ECONOMIC DEVELOPMENT

Developing countries are at a disadvantage in a world where timely delivery of products is important, because they tend to be far away from developed countries' markets. As already noted, this disadvantage may have decreased as technology has made it possible to deliver more rapidly over long distances but, at the same time, the disadvantage may have increased as the need for speed has risen with more rapidly changing technologies and tastes. All of this is straightforward, given what has been discussed above.

There is another issue, though, which has not been discussed and which must be important for developing countries: the factor intensity of time. Although having no hard evidence to support it, it seems obvious to the author that speed in production requires capital. That is, other things being equal, more rapid methods of producing goods are also more capital-intensive, in the sense of using a higher ratio of capital to labour. Humans are limited in the pace at which they can work, and beyond some limits simply adding more labour to a production process will not speed it up. Machines, on the other hand, often provide an increase in labour productivity precisely by letting a worker produce output in a shorter time. And as production processes become even more fully automated, the limits of human reaction times are left behind and production becomes faster and faster. This is just as true where capital substitutes for human brain power (computers) as where it substitutes for muscle power (heavy machinery). In all areas that the author has been able to think of, speed of production seems to be capital-intensive.

But we know in trade theory that factor intensity is very important for determining patterns of trade. The Heckscher-Ohlin model tells us that capital-abundant countries should export capital-intensive goods, while labour-abundant ones export labour-intensive goods. Since developing countries by definition are labour-abundant, this, together with the assumed capital intensity of time, means that developing countries tend to have comparative advantage in goods for which time is *not* a critical factor.

If true, this would not be a problem if the importance of time were fixed. Developing countries would simply specialize in non-time-intensive goods, and that would be that. They would still gain from trade and be able, one hopes, to harness those gains to the cause of their own economic development. But what if the importance of time is increasing, as suggested above? Then the capital intensity of time also means that developing countries find their comparative advantage in a range of products the demand for which is becoming more and more limited. This would suggest that the relative prices of the things that they can produce will go down over time, making them worse off.

Round Table 127: Time and Transport – ISBN 92-821-2330-8 – © ECMT, 2005

8. CONCLUSION

The author feels confident that our understanding of the importance of trade costs and, in particular, the time costs of trade, will expand rapidly over the coming years now that our attention has been directed to it. There was never, in fact, a good case for ignoring these costs, other than convenience, and as discussed above, the cost of that convenience in terms of misunderstanding of the world has become hard to ignore. Unfortunately, for the moment at least, it is equally hard to come to grips with trade costs in our economic models. Aside from a few obvious observations such as those offered here, we have so far little to offer to a world that needs to know how trade costs matter for trade. It is hard enough when the costs themselves are the straightforward resource costs of transporting goods from one country to another. When we try to grapple with the importance of the speed of that transportation and the costs of delay, we are on even less firm ground. Yet the payoff to doing so is great, and one looks forward to seeing a burgeoning body of research on this topic in the coming years.

NOTES

1. Consider *The Economist* magazine's Big Mac Index, which annually compares the US dollar prices of the Big Mac sandwich at McDonald's restaurants throughout the world. It finds, most recently in *The Economist* (2003), that these range from $1.05 in Uruguay to $5.79 in Iceland.

2. Of course, payment arrangements may in fact be different from this, with the buyer either prepaying with the order, on the one hand, or being given several months to pay after delivery on the other, for example. This does not change the fact of there being an interest cost associated with the time between shipment and delivery, but only determines who bears that cost.

3. For those costs that involve a purchase on the market, such as payment for transportation services, one might think that the provider of the service gains to the same extent that the user of the service loses, but that is not the case. All production involves real costs, in the sense of resources which are used for one purpose which could have been used for another. Therefore, the payments to service providers, although certainly desirable from a provider's perspective, represent costs to the economy, since it could have been using those resources to produce something else, something which could have contributed directly to economic welfare.

4. In the jargon of trade theory, costs of tariffs include only "triangles" of welfare loss, while the costs of real trade costs also include "rectangles" which are usually larger. Both of these terms refer to certain geometric areas which appear in the partial equilibrium supply and demand diagram used for analysing tariffs.

5. This is obvious for production stages which come after the delay. Suppose there are, say, five production stages with each taking inputs from the previous one, adding value and passing them along. Then a delay in delivering the output of stage three to stage four, say, will force the production line in stage four to shut down while it awaits its input. And then, even if it delivers its own output promptly to stage five, stage five will still have to wait the same time that stage four did. But these effects also pass back along the chain to earlier stages. Once stages four and five shut down to wait, their demand for inputs is reduced. Therefore stages one, two and three all need to produce less as well, perhaps by also stopping production for the same period of delay.

Round Table 127: Time and Transport – ISBN 92-821-2330-8 - © ECMT, 2005

BIBLIOGRAPHY

Anderson, James and Eric van Wincoop (2003), "Trade Costs," manuscript.

Deardorff, Alan V. (2003a), "Local Comparative Advantage: Trade Costs and the Pattern of Trade," manuscript, June 1.

Deardorff, Alan V. (2003b), "Time and Trade: The Role of Time in Determining the Structure and Effects of International Trade, with an Application to Japan," in: Robert M. Stern (ed.), *Analytical Studies in US-Japan International Economic Relations*, Cheltenham, UK and Northampton, MA: Edward Elgar Publishing Inc.

Evans, Carolyn L. and James Harrigan (2003), "Distance, Time, and Specialization," National Bureau of Economic Research, Working Paper No. w9729, May.

The Economist (2003), "McCurrencies" (April 24).

Krugman, Paul R. (1992), *Geography and Trade (Gaston Eyskens Lecture)*, Cambridge, MA: MIT Press.

Helliwell, John F. (1998), *How Much Do National Borders Matter?*, Washington, DC: The Brookings Institution.

Hummels, David (1999), "Have International Transportation Costs Declined?" manuscript, November.

Hummels, David (2001), "Time as a Trade Barrier," manuscript, July.

McCallum, John (1995), "National Borders Matter: Canada-US Regional Trade Patterns," *American Economic Review* 85 (June), pp. 615-623.

Pöyhönen, Pentti (1963), "A Tentative Model for the Volume of Trade Between Countries", *Weltwirtschaftliches Archiv* 90 (1), pp. 93-99.

Samuelson, Paul A. (1948), "International Trade and the Equalisation of Factor Prices", *Economic Journal* 58 (June), pp. 163-184.

Samuelson, Paul A. (1952), "The Transfer Problem and Transport Costs," *Economic Journal* 62 (June), pp. 278-304.

Tinbergen, Jan (1962), *Shaping the World Economy: Suggestions for an International Economic Policy*, New York: Twentieth Century Fund.

Trefler, Daniel (1993), "International Factor Price Differences: Leontief Was Right," *Journal of Political Economy* 101 (December), pp. 961-987.

Trefler, Daniel (1995), "The Case of the Missing Trade and Other Mysteries", *American Economic Review* 85 (December), pp. 1029-1046.

Turrini, Alessandro and Tanguy van Ypersele (2002), "Traders, Courts and the Home Bias Puzzle", paper presented at the 13th Congress of the International Economic Association, Lisbon, September.

Vanek, Jaroslav (1968), "The Factor Proportions Theory: The n-Factor Case", *Kyklos* 4 (October), pp. 749-756.

Vernon, Raymond (1966), "International Investment and International Trade in the Product Cycle", *Quarterly Journal of Economics* (May), pp. 190-207.

Yi, Kei-Mu (2003), "Can Vertical Specialization Explain the Growth of World Trade?", *Journal of Political Economy* 111, February, pp. 52-102.

TIME AND PASSENGER TRANSPORT

Yves CROZET
University Lumière Lyon 2
Laboratoire d'Economie des Transports
Lyons
France

TIME AND PASSENGER TRANSPORT

SUMMARY

Lyons, June 2003

INTRODUCTION

Time is money! Applying this common saying to their own sphere of activity, economists, following the example of Gary Becker, have for the past twenty or thirty years incorporated the issue of time into their analytical reasoning. A scarce resource, time is a topic of study that lends itself particularly well to demonstrations based on the principles of optimisation. Allocating time to the individual activities available when we decide upon our programmes of activity leads to varying levels of utility which rational economic man can choose between. Transport economists, following the work of C. Abraham and M.E. Beesley, have rapidly fallen into step with their generalist colleagues. They have done this by treating travel time as a component of an overall trip cost, termed the generalised cost.

Viewed from this standpoint, as we shall see in the first part of this paper, travel speed has become a key variable in overall transport demand, in that an increase in speed can potentially mean a lower generalised cost for the trip. By the same token, the choice of a given mode of transport will take account of relative speed, which will become determining in the modal split and which is closely correlated to the value of time for different categories of user. Transport users will thereby reveal preferences that public decisionmakers will have to take into account in investment in transport infrastructure. Economic logic will recommend that the priority for investment should be areas where time savings are possible, i.e. areas where a collective surplus can easily be achieved. Since one way to increase the collective surplus is to increase average travel speeds, economists have, for many years, argued in favour of introducing congestion charges, where the price paid by the user is designed to create fluidity and hence higher travel speeds that will reduce the time lost in travelling. This proposal has been adopted in many different areas where it has proved the robustness of its premise. However, in the case of urban congestion, problems have been encountered which, while they have not compromised the idea of road pricing, have nonetheless prompted the authorities to adopt a different approach to the relationship between time and transport.

The second part of this report considers the basis, contents and implications of this partial reappraisal, which can briefly be summed up as follows. If travel time, instead of being viewed as a variable that needs to be reduced, is treated as a constant in the activity programmes of individuals then, in certain specific contexts, public policies can be directed towards goals other than that of increasing the collective surplus through a trend increase in speeds. We shall note with interest that this viewpoint, like the preceding one, is based on a microeconomic analysis. This was notably the case of Y. Zahavi, who demonstrated the robustness and interest of a hypothetical double constancy, namely, that of the travel budget and that of the money budget relating to travel. Should the double-constancy hypothesis prove correct, it would have numerous implications for both inter-city and urban mobility. In both cases we are confronted with a very close link between economic growth and growth in the distances travelled. Indeed, this link is so close that, in pursuing the goal of sustainable mobility, we need to determine which modes should be given priority according to their relative energy consumption and emission of pollutants. This example provides a very rich illustration of the relationships between travel times and investment choices. In particular, what will be the coherence of collective choices regarding the value of time, on the one hand, and the discount rate on the other? This question particularly needs to be asked with regard to daily mobility. To counter the negative effects of the increase in the average distance of trips and the ensuing urban sprawl, should we consider another way of charging for trips made in urban and outlying areas? Should the idea that the impact of road pricing is compensated for by the increase in the average speed of trips be replaced, as implicitly recommended by many urban transport policymakers, by that of "charging for generalised cost" in order to increase the two components of this cost, namely, price and time?

1. TIME: A KEY VARIABLE IN INDIVIDUAL AND COLLECTIVE CHOICES

Since transport is considered as intermediate consumption, transport demand from individuals is usually treated as derived demand. In other words, for travellers, transport is a necessary but not a sufficient condition for the performance of our various activities. From this standpoint, economic analysis assumes that individuals will seek to reduce the cost of transport, primarily by increasing travel speeds. The desire for speed increases commensurately with the increase in the value of time, which itself is closely linked to higher incomes. Linking this trend increase in the value of time to another means of measuring the price of time, namely, the discount rate, reveals another dimension to the relationship between time and transport -- one which, by aggregating individual demands in the economic calculation, seeks to assist the decisionmaking process, notably with regard to the strategic issue of transport infrastructure programming. Discounting, i.e. calculating, for a given date, values estimated to obtain in the future, is one way in which to integrate the relationship between time and transport into a collective and intertemporal approach. It is currently a prerequisite for understanding the priorities of public policies -- notably, the reasons for which policies are directed towards the development of transport systems which afford substantial time savings (air transport, high-speed rail, inter-city motorways) and justify the introduction of road pricing, which takes account of the improvement to service quality offered by higher speeds. But is the pairing of speed and road pricing consequently going to become the basis for public policies in all areas where an ability to pay and congestion exist? This is by no means certain, and in this particular area there is undoubtedly no universal answer, as may be seen in the limited use made of urban tolls, despite the fact that they make perfect sense from an economic standpoint.

1.1. Time: a scarce resource with a monetary equivalent

In a famous article published in 1965, G. Becker proposed a general analysis of the allocation of this scarce resource by linking it directly to the monetary components of consumers' choices. By integrating his analysis into the new consumer theory, G. Becker suggested that the utility of an individual did not derive solely from the quantity of goods and services consumed but also from the commodities to which they corresponded (meals, childcare, personal care, golf matches, evenings at the cinema, etc.). The consumer is therefore not a passive being, in reality he is the producer of the commodities he consumes. The production of these commodities requires not only goods and services but also time. These two inputs must therefore be related to the two types of allocation that individuals can make, namely, the allocation of time and the allocation of money. These two allocations are therefore closely linked, since it is possible to increase the allocation of money by modifying the use we make of our overall allocation of time.

This form of reasoning leads us to a classical problem of optimisation. The individual must maximise his utility (U) by combining the inputs X (goods and services) and T (time) which he needs to produce commodities (Z). Commodities number from 1 to n and are indexed i.

Total utility therefore depends on various commodities:

$$U = U(Z_1,...,Z_n)$$

The combination of goods and time must be taken into account for each commodity i:

$$Z_i = f_i i(x_i,t_i)$$

In view of the budget constraint [P stands for prices, W (Z_n) wages, V unearned income],

$$\sum P_i X_i = W (Z_n) + V$$

and the time constraint (T = total time available),

$$\sum T_i = T$$

maximising well-being will lead to an optimum allocation of activities that will make the marginal utility of each activity equal to its shadow price, weighted by the marginal utility of income (where λ is the marginal value of income and Π_i the shadow price of activity i), i.e.:

$$\delta U/\delta Z_i = \lambda \Pi_i$$

In all, the optimum combination of inputs for each activity is (where τ is the shadow price of time):

$$(\delta Z_i/\delta T_i)/(\delta Z_i/\delta X_i) = \tau/P_i$$

which means that the relative allocation of time and goods to an activity must correspond to the ratio of the price of time to that of good i.

Choices will therefore depend upon the relative price of goods and services, but also and above all on the wage rate, that is to say, the relative incentive to trade time for income. One of the main conclusions of the work of G. Becker is to show that an increase in the wage rate, or employment opportunities, significantly modifies programmes of activity and, notably, the choices made by women. As soon as the latter are able to obtain a wage outside the home, they will substitute working time for time in the house, which will be replaced by the purchase of goods (washing machine, pre-cooked meals) or services (child-minding, domestic help) that will make this substitution easier to achieve. Reducing the number of children is obviously another way of reducing time spent on domestic tasks. As if to demonstrate the strength of G. Becker's argument, this substitution phenomenon has been noted in all countries experiencing a certain level of development and economic growth. Female employment, lower fertility levels and an increase in the number of single and divorced people are therefore some of the consequences, among others, of the impacts of higher income levels on the allocation of time.

The transport field is also concerned by the substitutability of various approaches to time management. To the extent that transport is primarily an intermediate consumption, in the form of a derived demand related to production of a specific "commodity" (work, recreation, etc.), it will be tempting to reduce the time devoted to this intermediate consumption. In the same way that goods or service can be substituted for time spent on domestic tasks in order to be able to use that time for work, so too can a rapid mode of transport be beneficially substituted for a slow mode. The resultant time saving can then be reinvested in recreational activities and/or work, particularly if the latter affords access to a higher income, one of whose possible uses will be to pay for the increased travel speed[1]. The seminal work of C. Abraham and M.E. Beesley in this area clearly showed that time had a very real value in the transport field. According to their income, preferences and the opportunities for activity and transport open to them, etc., individuals are willing to pay a certain price to gain access to a faster mode. The figure below demonstrates this in the long run. As income levels rise, slow transport modes give way to faster modes which significantly increase the opportunities to diversify our programmes of activity, the basis of the desired increase in utility.

Figure 1. **Trend in distances travelled by person since 1800 in the United States**

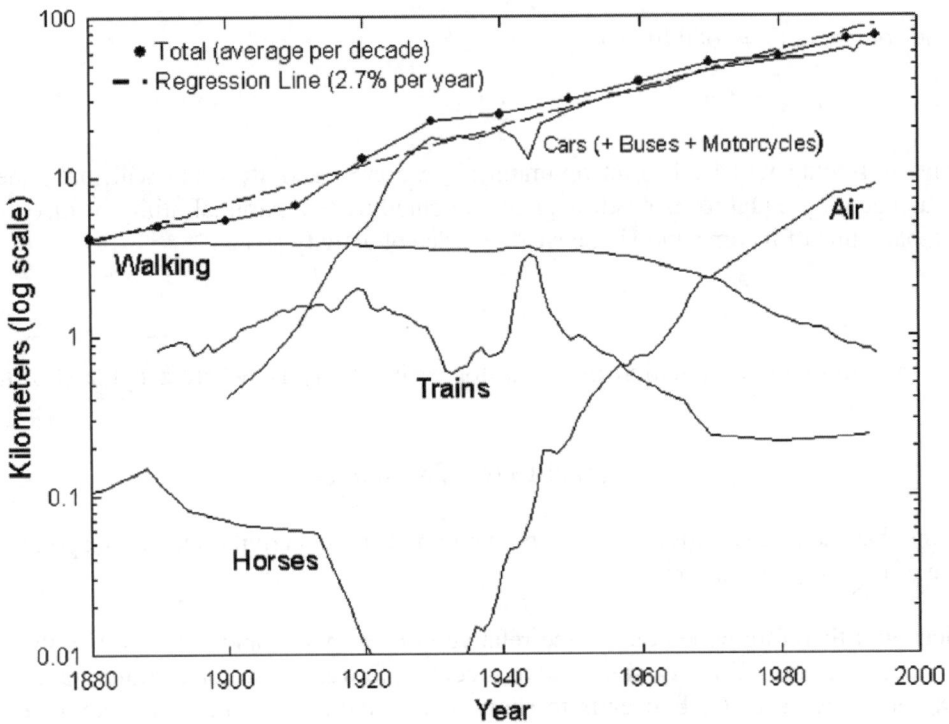

Source: Ausubel, J.H., C. Marchetti, P.S. Meyer.

On observing the main trends, reported here in the United States, it can be noted that with economic growth and increased real incomes there is more than simply a substitution effect at work. Slow modes (walking, horses, etc.) are replaced by faster modes but, at the same time, there is an increase in the average distances travelled annually (trend increase of 2.7 per cent). There is, therefore, an income effect in the form of an overall increase in transport demand, a phenomenon that is directly linked to the substitution of slow modes by fast modes. We find ourselves confronted here with the same paradox identified, on the basis of the work of Garry Becker, in relation to leisure time. When wage rates and employment opportunities increase, leisure time must logically be converted into work time, which represents access to new opportunities for our programmes of activity.

Round Table 127: Time and Transport – ISBN 92-821-2330-8 - © ECMT, 2005

In reality, this only obtains if leisure time is a kind of "time-out", what economics would call an inferior good, whose consumption tends to decline when income rises. If, on the other hand, leisure time is treated as prime time, and therefore a normal good (whose consumption increases in line with income) or even as a superior good (whose consumption increases at a faster pace than income), then we may find both higher incomes and increased leisure time. The same reasoning applies to the transport field. As income levels rise, some types of trip may be reduced because they correspond to an inferior form of transport service. Others, in contrast, will rapidly expand, notably fast modes, as we shall demonstrate below by examining the key role of the generalised trip cost.

1.2. Time and generalised travel cost: Towards a composite monetary equivalent

One way to illustrate the key role of time in the transport field is to examine the relationship between the speed of a transport mode and the resultant volume of traffic. In this area, the gravity model, whereby the volume of traffic between two urban areas can be estimated on the basis of their relative weight (i.e. population) and the distance between them, can be taken as a fundamental given. Even though there are some exceptions to the general line of argument, notably when two urban areas are separated by an international border, gravity models have a good capacity for predicting potential traffic flows. However, in measuring the distance between two urban areas, what matters above all is transport time in relation to its cost rather than the actual number of kilometres. The term "gravity" is therefore explained by the fact that traffic is proportional to the populations of the two interconnected centres and inversely proportional to the generalised costs. Distance is therefore measured by the generalised cost of the trip, which may be expressed as follows:

$$C_g = p + hT_g$$

where:

P = Monetary price of the trip between location i and location j;
Tg = Generalised time between i and j;
H = Monetary parameter representing the average value of time in the eyes of travellers.

It is worth noting that the generalised cost takes account of monetary cost, the full transport time and a term relating to the way in which this transport time is perceived. The aim here is to take account of load breaks, the frequency of services in the case of public transport, number of transfers, etc. There is therefore a qualitative dimension to placing a value on travel time. To take this qualitative dimension into account, according to the mode studied, the Tg parameter can be constructed in greater detail in order to take better account of not only travel time but also the access time upstream and downstream, if necessary, as well as the intrinsic performance and quality levels of the mode in question.

In the case of a trip by rail or by air, for example, the following parameters could be taken into account:

- Travel time in the form of average travelling time between the origin and destination points in zones i and j;
- An indicator of the average interval between two trains (planes) according to the hourly schedule of a day's service;
- The number of transfers (train or plane) the traveller is required to make (load break);
- Frequency of trains or planes on the route;
- A constant representing times at journey ends.

This will provide an aggregate total time, a physical value that must be matched to the price of the trip by choosing an average time value for passengers.

- To measure this parameter, from a theoretical standpoint, the economic analysis is based on the principle that time is scarce. The individual chooses between the various activities possible by comparing the utility he derives from an activity and the share of the total time available to him which is thereby consumed. Consequently, time spent travelling is time taken away from other activities.

- From a practical standpoint, the placing of monetary value on time occurs through the notion of Time Value (TV), or the monetary value of time. This value is established through the study of individual behaviour patterns and may therefore be viewed as a behavioural value: individuals' willingness to pay in order to save time.

The value of time is usually obtained through direct methods of evaluating effects, notably through stated preference surveys or revealed preference methods. The studies which have been conducted, even though they still contain many biases, have provided us with greater insights. Nonetheless, it still remains extremely difficult to distinguish fine classes of uniform individuals in terms of time value, just as it would be equally difficult to segment the customers of a transport route into subclasses and estimate the number of users and the forecast changes in each subclass. Instead, an average approach is used with regard to the opportunity cost of time, by relating the hourly value of time to average hourly salaries.

The value of time, which symbolises the monetary value placed on time by individuals, is therefore the product of a process of simplification. As such, it theoretically depends upon individual social economic factors such as wealth, income, socioprofessional category, reason for trip, mode, etc. It is also common practice to choose different time values in urban areas for inter-city trips, as suggested with regard to France by the Boiteux 2 report, which recommends that a distinction be drawn between three types of reason and that better account be taken of comfort through the introduction of discomfort costs of 1.5 for congested travel conditions (+50 per cent of the cost of time spent travelling) and 2 for waiting times (+100 per cent of the cost of time) in assessments.

For inter-city travel, the time values proposed differ according to mode in order to take account of differences between customers. The time value is, on average, higher in a plane than in a train or a car. The values proposed therefore distinguish between the distance travelled and, in the case of rail transport, class of service (see Tables 1 and 2).

Table 1[2]. **Time value proposed per passenger in urban areas (in 1998 euros per hour)**

Mode of travel	% of wage cost	% of gross salary	France as a whole	Ile-de-France region
Professional trip	61%	85%	10.5 €	13.0
Journey-to-work	55%	77%	9.5	11.6
Other trips (shopping, leisure, tourism, etc.)	30%	42%	5.2	6.4
When no details are available of the breakdown of traffic by trip reason, average value[3]	42%	59%	7.2	8.8

Round Table 127: Time and Transport – ISBN 92-821-2330-8 - © ECMT, 2005

Table 2. **1998 time value per passenger in urban areas (in 1998 euros per hour)**

Mode	For distances below		For distances d between 50 km or 150 km and 400 km	Stabilisation for d > 400km
	50 km	150 km		
Road	8.4 €	-	50 km<d TV = (d/10+50).1/6.56	13.7
2nd class rail	-	10.7	150 km<d TV = 1/7(3d/10+445).1/6.56	12.3
1st class rail	-	27.4	150 km<d TV = 1/7(9d/10+1125).1/6.56	32.3
Air			45.7	45.7

Because of economic trends in prices, the consumption of main items, income and wealth, the value of time will also vary. The rule for the trend in time value recommended in the Boiteux 2 report is that of a trend in per capita household consumption with an elasticity of 0.7.

On the whole, even though stated preference surveys have been conducted, it is extremely rare that a person is capable of specifying a monetary equivalent for the value of his own time at a given time and for a given activity. The value of time is, in most cases, revealed through behaviours that, in fact, are not quite as rational as analysis might suggest. C. Segonne, for example, has shown how the users of the Prado-Carénage Tunnel in Marseilles overestimate the time saved through use of this toll infrastructure. Symmetrically, non-users overestimate the time they would have saved by taking this new route. However, the fact that the rationale in the field of transport, as in other areas of consumer choice, is limited by the impossibility of having detailed knowledge of the alternatives, does not invalidate the thrust of the analysis: the statements or behaviours of individuals reveal that the latter place a certain value on time. The limits to rationality may result in routine or gregarious forms of behaviour which weaken the capacity for permanent optimisation accorded to actors at the microeconomic level. In no way do they invalidate the notion that individuals try to reduce the costs relating to performance of a given activity, notably by taking account of the associated travel time.

On this basis of a generalised preference for speed, what the transport economist will seek to determine is the elasticity of demand to the generalised cost. Depending upon the type of transport, and more specifically the type of activity relating to that transport, demand for transport will respond more or less strongly to a variation in price. If the "transport" good is a superior good -- that is to say, a good exhibiting high utility -- it is most likely that the elasticity of demand to price will also be high. Any decrease in the generalised cost of transport, made possible by higher speeds and/or lower monetary costs, will fuel strong growth in demand. Thus, if we return to the gravity model, we can say that the volume of traffic between two zones i and j will be expressed as follows:

$$T_{ij} = K \frac{P_i P_j}{C_{g_{ij}}^{\gamma}}$$

where:

P_i and P_j = Respective population of the two geographical zones i and j;
Cg_{ij} = Generalised cost of the transport in question between zones i and j;
γ = Elasticity of traffic to the generalised cost;
K = Adjustment parameter.

The numerator contains the factors of attraction and the denominator the factors of repulsion or resistance, whose values will increase commensurately with elasticity. It is for this reason that some analysts sometimes suggest that the law of gravity applies even more strongly for values of elasticity γ higher than 1 or even around 2. A high elasticity in this instance means high sensitivity to the decrease in generalised costs and, in particular, the reduction in travel time afforded by higher speeds. Since speeds do not rise at the same rate in different modes, it is important to note that, apart from the questions of elasticity, it is necessary to show how the change in relative speeds has a quite significant impact on relative generalised costs and hence the modal split of traffic flows.

1.3. Price-time models and modal choice

To illustrate the key role played by relative speeds, we shall briefly discuss the rationale that has allowed France to make high-speed rail projects an economically credible proposition. Use was made of an econometric model designed to explain the modal split between rail and air for new high-speed railway lines; this model combines a price-time model with a gravity model and therefore takes account of two modes of transport (rail and air).

The first step is to express the generalised costs associated with each of the competing modes of transport, since the model is based on the assumption that a passenger will choose between the two modes according to the value he places on his time and the characteristics of the costs and travel time of each mode. User k will therefore choose the mode with the lowest generalised cost once his time value, hk, has been taken into account.

Let us assume that we are modelling a modal split between rail and air. The respective prices of rail and air are therefore P_F and P_A; T_F and T_A are the respective journey times (including final legs), and the generalised costs for user k are expressed as follows:

$$Cg^k_A = P_A + h_k T_A$$
$$Cg^k_F = P_F + h_k T_F$$

On a given route i, there is a time value h^i_0 such that:

$$Cg_A = Cg_F$$

which is known as the time indifference value on route i. If h_k is less than h^i_0, user k will choose rail, or failing that, air travel.

It is assumed that the passenger population on a given route is characterised by a passenger time value g(h) distribution whose function is:

$$F(h) = \int_0^h f(x)\, dx$$

This gives the proportion of trips whose time value is less than h.

Accordingly, the proportion Y_i of air users in total traffic will be given by:

$$Y_i = \int_{h_i}^{+\infty} f(x)\, dx = 1 - F(h_i)$$

Round Table 127: Time and Transport – ISBN 92-821-2330-8 - © ECMT, 2005

This is illustrated in the two figures below:

Figure 2. **Comparative generalised costs of rail and air**

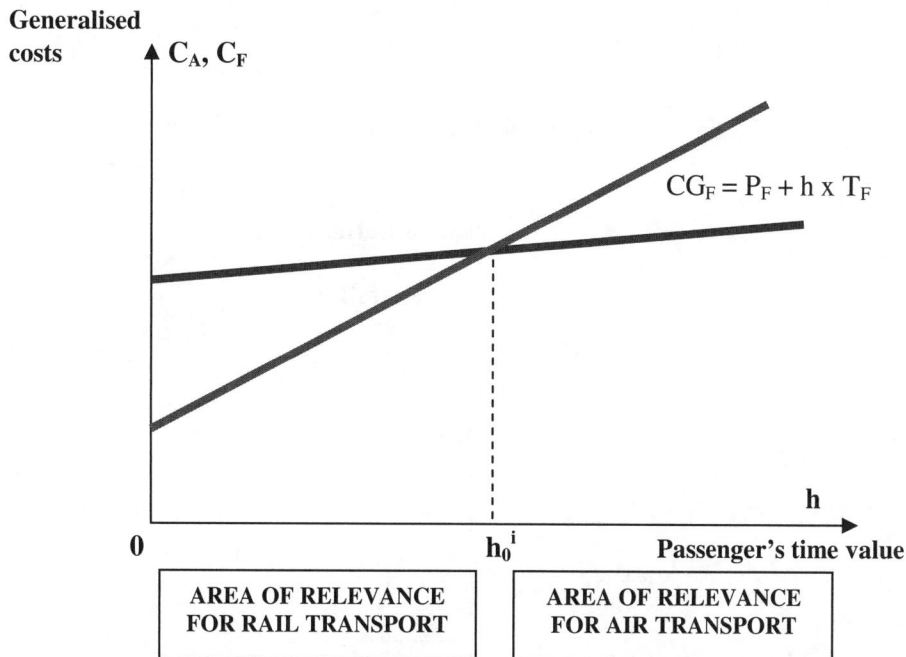

If we now put in place a high-speed train allowing substantial time savings, this will modify the generalised cost of rail transport, all things being equal. The gradient of line Cg_F now will shift.

Figure 3. **Improvement of the market share of rail as a result of the introduction of high-speed services**

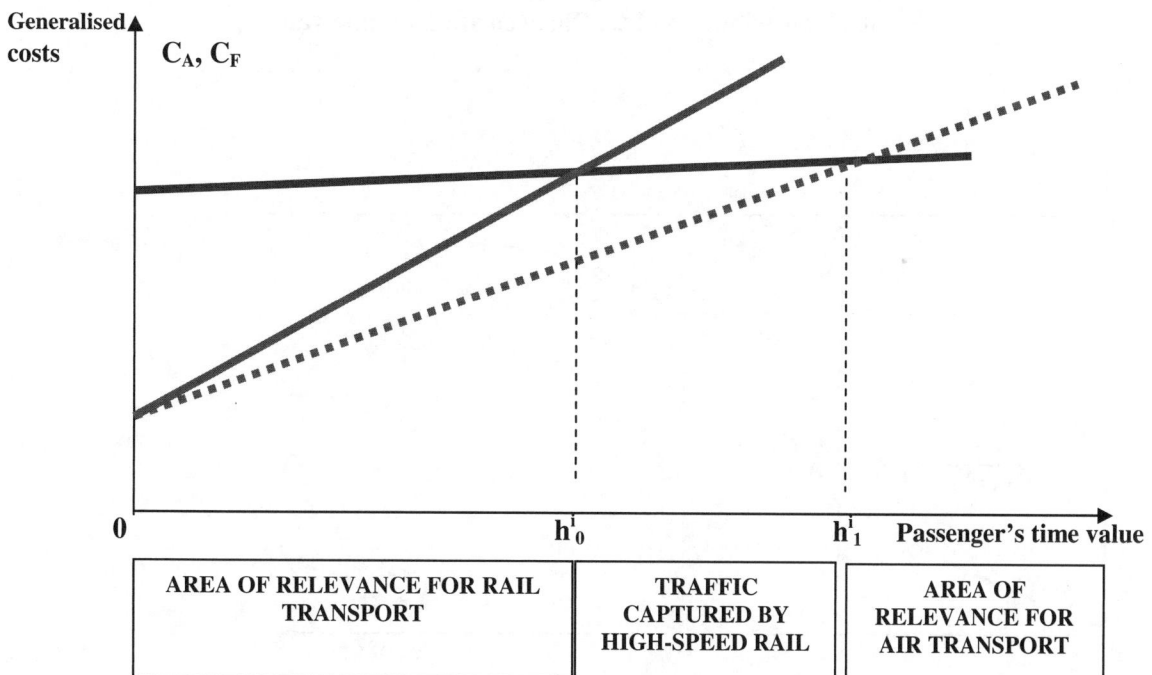

where:

Train Cg_F $\quad = P_F + h \times T_F$

Air Cg_A $\quad = P_A + h \times T_A$

TGV Cg_{tgv} $\quad = P_{tgv} + h \times T_{tgv}$

Figure 4. **Distribution of time values**

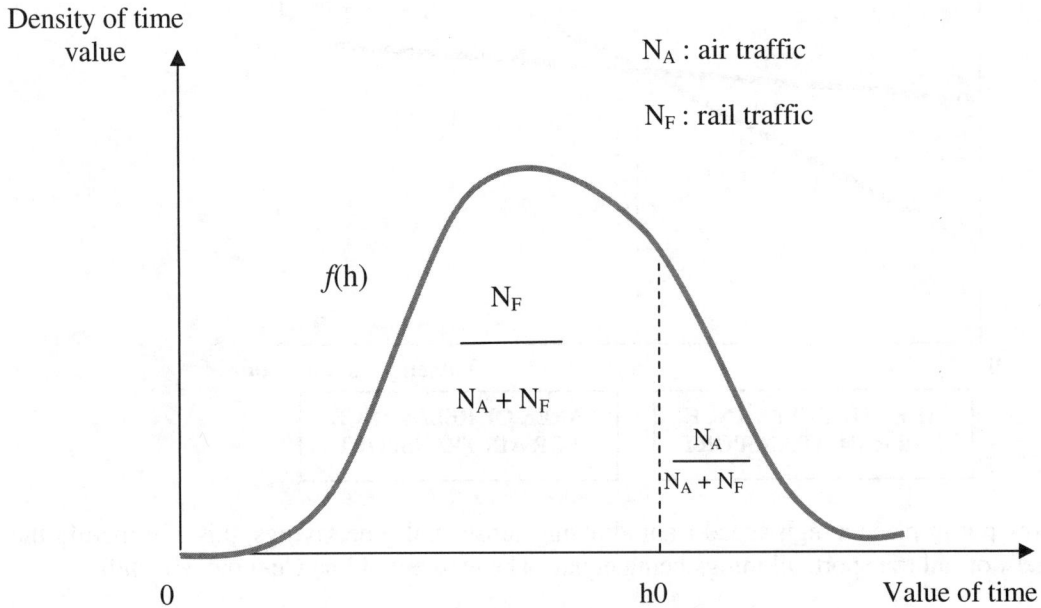

Density of time value

N_A : air traffic

N_F : rail traffic

$f(h)$

$$\dfrac{N_F}{N_A + N_F}$$

$$\dfrac{N_A}{N_A + N_F}$$

0

h0

Value of time

Figure 5. **Distribution of traffic according to time value**

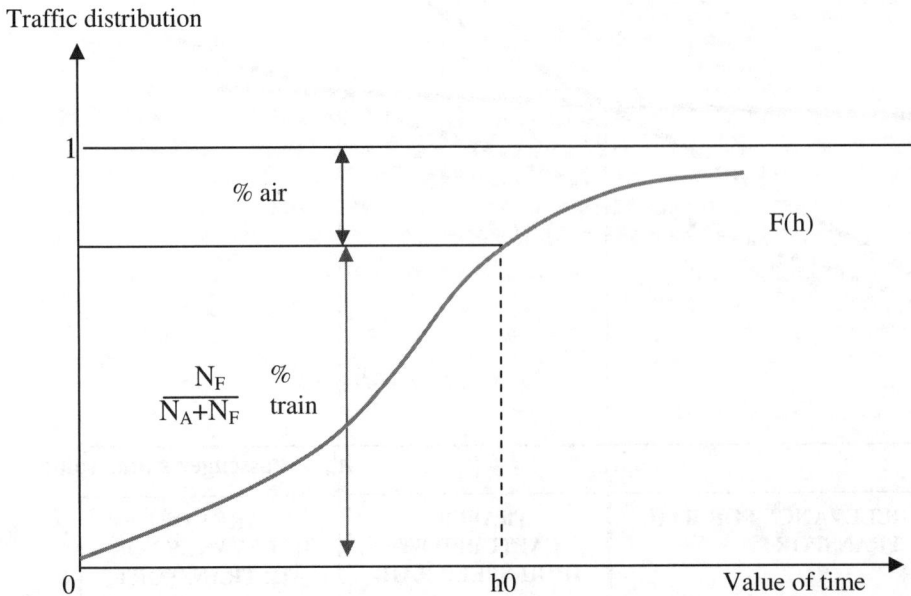

Traffic distribution

1

% air

$\dfrac{N_F}{N_A+N_F}$ % train

F(h)

0

h0

Value of time

Round Table 127: Time and Transport – ISBN 92-821-2330-8 - © ECMT, 2005

Because of the wealth of data available regarding income distribution among the population in a large number of countries, we can formulate a time value density function for log-normal $f(h)$, that is to say:

$$f(h) = \frac{1}{h\sigma\sqrt{2\pi}}\exp.-\left(\frac{(Logh - Logm)^2}{2\sigma^2}\right)$$

where σ is the standard logarithmic deviation for time values and m the average time value.

Adjustment of the model consists in calibrating the log-normal law parameters, that is to say, the standard logarithmic deviation of time values and the average time value. The calibration must apply to as many routes as possible on which the two transport modes are competing (in this case rail and air). This desired spread in the data collected makes it possible to ensure that the stability of the parameter adjustment is properly verified and, in particular, that a correlation does actually exist in the country concerned between the average time value expressed in constant terms and the volume consumption of households.

This model is aggregated to the extent that it reconstitutes market shares. Once the model has been calibrated, the time value is set at a specific value in order to be able to test various trend scenarios in the transport system. To obtain medium- and long-term projections, the future time value is correlated with the forecast increase in income. It should be noted that there are several variants of the price-time model. What in fact differs from one model to another is the way in which the generalised cost is formulated. This comment suffices to show that the time value is not self-evident in the eyes of the economist; on the contrary, it is a value construction on the basis of what are frequently highly sophisticated lines of reasoning.

Recent studies (De Palma and Fontan, 2001; Bayac and Causse, 2002; Hensher, 2001) have shown that it was possible to improve the account taken of time value in transport demand models. The basic idea put forward in these studies, based on disaggregated models integrating a random dimension into individual demand, is to consider a non-linear relationship between time savings and utility. Their main outcome is not a reduction but rather an increase in the revealed value of transport time and therefore a *de facto* greater preference for speed.

1.4. Time and choice of investment

Time saving is therefore one of the main goals of investment in the transport field. It is essential to take account of time savings when calculating the social and economic viability of a project, that is to say, its viability after account is taken of the collective, and not simply the financial, interest of the project. When extended to the non-monetary aspects of the surplus afforded to the community by new transport infrastructure, cost-benefit analysis focuses on time savings, which usually account for four-fifths of the non-monetary benefits.

1.4.1. Discounting: another economic approach to time

In economics, discounting allows account to be taken of the time dimension and to compare sums at discrete horizons. The discounting principle is analogous to the principle of capitalisation. If a capital C_0 is placed on a market at an interest rate r in year 0, the capital in year n will be C_n where:

$$C_n = C_0 * (1 + r)^n$$

The discounting procedure uses a **discounting rate** *a* (usually per annum), representing the preferences for the availability of money over time. One euro available in a year is equivalent to 1 + a euros available today. A sum of S_n in year n is only taken into account as S_0 where:

$$S_0 = S_n/(1 + a)^n$$

The discounting rate, and its level, therefore translate a more or less significant preference for the present. An individual whose preference is for the present will have a very high discounting rate, which will severely penalise future goods. Conversely, an individual who gives priority to the long term will have a low discounting rate. It would even be feasible for individuals to have a preference for the future rather than the present, in which case they would have a negative discounting rate.

The first economic studies on discounting are now quite old. Ramsey (1928), Evans (1930) and Hotelling (1931, as part of his work on natural resources), made major contributions, as well as Massé (1946) or Arrow and Kurtz (1970, on the role of public capital). In the light of this work, it can only be said that the discounting rate represents more than simply a more or less strong preference for the present.

– In the first analysis, the process of discounting may be justified by what economists refer to as the "pure" preference for the present, which in some respects translates the "impatience" of economic agents. An impatient consumer will have a high discounting rate and the immediate consumption of the product provides him with greater satisfaction than consumption at a later time. This pure preference rate, denoted *p*, is obviously a behavioural value. Arrow (1995 and 1996) proposes, on the basis of ethical and empirical conclusions, to adopt a pure preference rate of 1 per cent. Cline (1999) proposes, for his part, a rate of the order of 2 per cent. Other authors recommend adopting a rate of 0, particularly with regard to moral considerations regarding future generations when projects have intergenerational effects. Proposals may therefore specify different rates according to the length of the calculation. For investments of less than thirty years, the pure preference rate for the present (approximately 2 per cent) may be taken into account in the discounting process. For periods of more than thirty years, it must be taken as 0. Other authors such as Harvey, Heal, Overton and MacFadyen have proposed various formulae for the discounting rate, introducing a variation in the latter over time[4].

– The discounting rate also translates a "wealth effect". Future generations will be wealthier than the present generation. As a result, the utility of consuming a euro today is higher than that which will be derived from consuming a euro in several years' time, even at a zero rate of inflation. This means that the utility of a euro for a "poor" individual is greater than that derived by a "rich" individual from the consumption of the same sum of money. In this respect, the discounting rate must not be confused with the rate of inflation, since we are talking here in terms of constant money. The discounting process therefore does not translate the effect of depreciation of the value of money as a result of an increase in consumer prices. This wealth effect, denoted as $\theta*g$, is often related to economic growth. θ is therefore a term that takes account of the marginal utility of income (the value usually assigned is close to 1.5, which approximately matches the inverse elasticity of the marginal utility of income) and *g* is the per capita GDP growth rate (the long-term growth rates usually adopted range from 2 to 4 per cent).

– A third approach to discounting, which embraces the two preceding ones, is that of "opportunity cost". Since the immobilisation of capital is unproductive, the discounting rate can take account of the potential profit to be earned from an alternative use of a given capital, such as investing the capital in a financial market offering a guaranteed return. This justification of the discounting process also takes account of the constraint of scarce financial resources. This opportunity cost of money, generally denoted r, may be related to the real rate of interest in the financial market (the rate adopted may be the actual rate applicable to long-term obligations, i.e. a value of around 4 per cent).

In sum, the discounting rate, as a method of integrating time into economic calculations, may be defined as the aggregation of three terms: the "pure" preference of economic agents for the present at a rate p; the "wealth effect" $\theta*g$, generated by the action of time; and the "monetary opportunity cost" r. However, such an aggregation does not simply mean adding the three terms, rate p, $\theta*g$ and r, together. Böhm-Bawerk, followed by Cline[5], suggest that a social discounting rate must be adopted for individuals (or a "pure" discounting rate) to take account of p and $\theta*g$, i.e. the pure preference for the present and the wealth effect. This rate would not take account of the opportunity cost of money. In contrast, for firms and public investors, it is the opportunity cost which prevails in the determination of the discounting rate. The pure preference for the present and the wealth effect are implicitly part of a wider notion, which is the opportunity cost of money.

Figure 6. **Discounting rate: three approaches to the value of time**

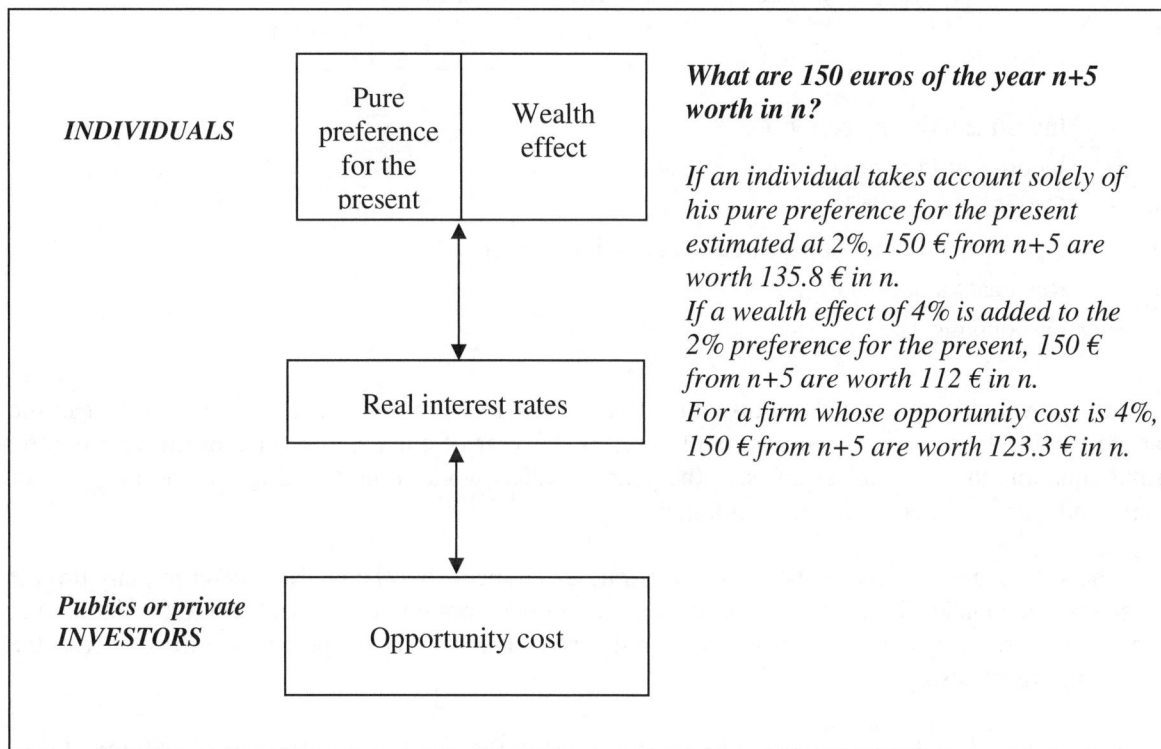

More than a simple economic calculation, specifying the discounting rate to apply calls for specification of the framework for and context in which projects are implemented. For example, the nature of the process of economic globalisation, and to an even greater extent the globalisation of capital, challenges the methods currently used to determine discounting rates (see Lind, 1990; Obstfeld, 1986). The same applies to the account taken of strategic elements; indeed, to such an extent that the choice of discounting rate for major infrastructure projects, while admittedly based on economic rationales, ultimately amounts to a political choice, particularly in the area of transport infrastructure.

1.4.2. *Time and public economic calculation in the field of transport infrastructure*

The discounting rate is extremely important in public economic calculations, since its level will partly determine the approach adopted by government to investment. The integration of time is not only a feature of the discounting rate, it is also a component of the time savings afforded by increased travel speeds. It can be seen in the equations below, which classically specify the discounted cash flow and the internal rate of return (IRR) in economic and social terms.

The discounted cash flow is the other facet of the Net Present Value (NPV), but takes account of the interest to the community in estimating the monetary value of the various costs and benefits of a public investment. The benefits estimated in monetary terms, denoted Aj, are decisive, in the numerator of the equation below, for determination of the socioeconomic viability of a project.

$$BNA = \sum_{j=t_p-t_r}^{j=t_n-t_r} \frac{-\Delta I_j + \Delta R_j - \Delta C_j + \Delta A_j}{(1+a)^j} + \frac{K_{t_n}}{(1+a)^{t_n-t_r}}$$

Ij	=	Investment during period J;
Rj	=	Income during period J;
Cj	=	Costs during period J;
Aj	=	Non-monetary but monetarised benefits during period J;
Kt	=	Residual value;
a	=	Discounting rate.

All the variables presented are discounted values for a reference year, t_r. For investment in the transport field, the reference year commonly used is the year of entry into service or the year before. In the equation above, t_p corresponds to the year in which work commenced and t_n the last year of operation taken into account in the calculation.

The socioeconomic IRR is the discounting rate that cancels out the NDB. Therefore, any project whose socioeconomic IRR is higher than the reference discounting rate is considered to be viable, that is to say, the project generates a sufficient overall surplus (i.e. its NDB is positive) compared with the initial investment costs.

The way in which time is taken into account is therefore decisive in two respects with regard to the public economic calculation:

– Firstly, the choice of a specific time value, representing a more or less high percentage of the average wage, will significantly modify the outcome of the project; particularly in view of the fact that the time saving often accounts for a non-negligible share of the factors which make the numerator positive in the first term of the equation above.

– Secondly, the choice of discounting rate, if high, can penalise the implementation of long-term projects by encouraging investments offering a speedier return for the community. The reference rate can therefore be interpreted as the community's preference for projects whose impacts are rapidly visible or, conversely, for higher capitalisation of the long-term effects.

By way of an illustration, let us consider the following, albeit somewhat simplistic, example. The authorities wish to improve the link between two major economic centres. They have two possible choices of investment: they can either improve the road link or improve the public transport link, both of which offer similar time savings for the same investment costs.

If the priority is given to short-term considerations with a high discounting rate, the investment will tend to be in favour of an improved road link in that the initial time savings will benefit a much greater number of users (assuming that road is used more than public transport in the initial situation). Even if the external costs rise (pollution, noise, etc.), the surplus generated during the first few years will be higher than that afforded by the public transport project. The longer term may see the emergence of unwanted effects, such as growth in car ownership, changes in the travel patterns of individuals -- who take advantage of the new infrastructure to live further away and therefore to travel longer distances, a decline in the share of public transport and increased congestion and hence pollution. With a high discounting rate, all of these effects (provided that they can be predicted and estimated) are relatively unimportant at the time of the calculation and have merely a limited impact on the rate of return on the road investment project. The road project will therefore remain superior to the projected investment in public transport. In contrast, with a low discounting rate, these long-term effects can have a significant impact on the calculation of viability and can lower the socioeconomic IRR of the road project to a level below that of the public transport investment project, particularly if the external costs (poor safety levels, pollution) relating specifically to the road projects are evaluated at high levels.

To a certain extent, account can be taken of time in a variety of more or less conflicting ways at the very heart of the economic calculation. If a high time value and high discounting rate are chosen, a strong preference for the present and for speed emerges. A modest, if not zero, discounting rate and low time value, on the other hand, work in favour of a preference for the future and low inclination for speed. However, it is also possible to adopt scenarios which partially work against a high time value and low discounting rate or, conversely, a low time value and high discounting rate. In this respect it is instructive, for a given discounting rate, to observe the way in which collective judgements are implicitly made in favour of or against a given time value. When faced with congestion problems in particular, does the objective of increasing speeds remain relevant at all times and under all circumstances?

1.5. Congestion charging: illustration or cornerstone of the importance of the travel time variable?

In 1930, when A.C. Pigou presented his famous example of internalising congestion costs, including time losses, through the introduction of a charge, his main objective was to achieve maximum fluidity through the optimum allocation of traffic flows to two competing routes. This approach amounted to establishing an explicit link between engineering and economics, which his successors were to achieve more precisely by making use of the insight provided by the speed-flow curve. Building on this technical foundation, economists subsequently developed their own set of problems by focusing too exclusively on the relationship between charging and infrastructure financing, which reveals the preference for speed. However, in urban areas, this preference faces some major obstacles.

1.5.1. The speed-flow curve and the knowledge gained from traffic engineering

The "standard" static congestion model is a relatively straightforward construction. A given infrastructure is considered to have constant capacity with a single input and single output. The "fundamental" diagram derived from traffic engineering data describes how speed V (measured, for example, in metres/second or kilometres/hour) declines as density D (measured in terms of the number of vehicles per metre of carriageway) increases. The maximum speed, V_{max}, is assumed to be reached for a positive density (which explains the flat portion at the beginning of the curve, and that the maximum density D_{max} corresponds to zero speed (the flow of vehicles becomes a "stock" of vehicles). By analogy with the theory of fluid dynamics, the vehicle flow, F (measured in terms of the number of vehicles per second), is the sum of D and V. Consequently, there is a maximum flow, F_{max}, corresponding to a given combination of speed and density, denoted $V^{\#}$ and $D^{\#6}$, respectively.

Figure 7. **Speed and traffic density**

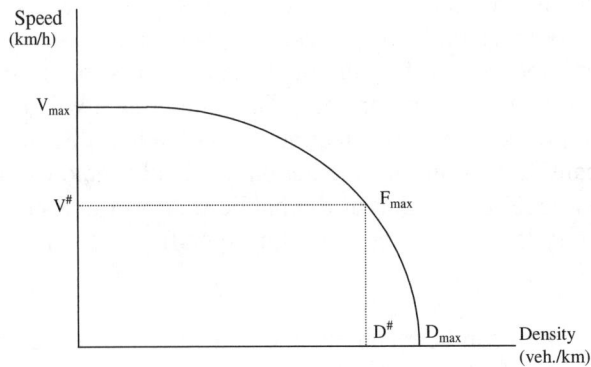

This speed-density curve can be used to derive a speed-flow curve exhibiting a characteristic profile: there is a positive correlation between flow and speed F_{max}, after which the correlation is negative (a flow higher than F_{max} implies a speed lower than $V^{\#}$.

Figure 8. **Speed-flow curve**

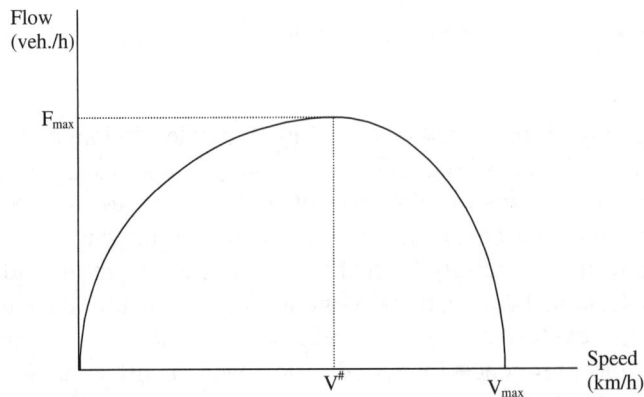

If it is assumed that solely time costs determine the cost of the trip made by users, and since travel time is known to be inversely proportionate to speed, it is possible to use the speed-flow curve to derive a curve representing the average cost of the trip for a given distance and time value. This curve is the average cost-flow curve which provides the basis for static models:

Figure 9. **Relationship between flow and the cost of the trip in terms of time**

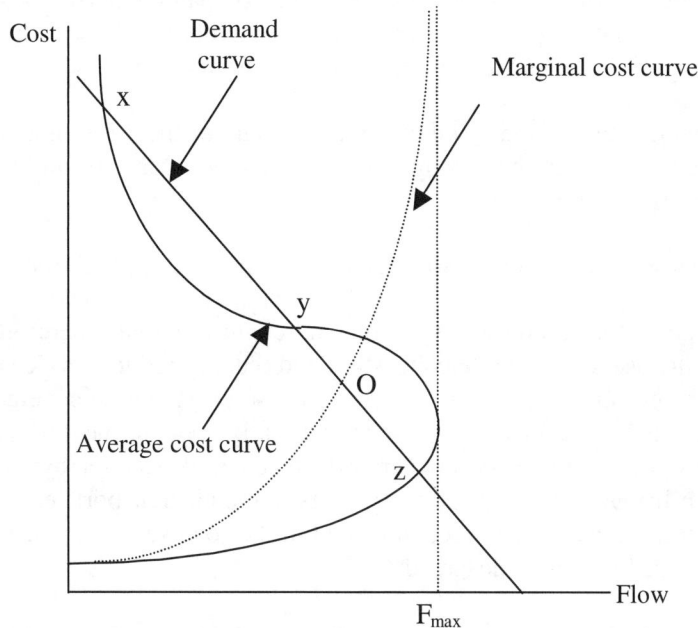

The maximum flow rate, F_{max}, is based on the assumption that there is a certain amount of congestion. The gradient of the average cost curve in the lower section of the curve is due to the fact that if the number of motorists increases, speed decreases but flow continues to rise. Above maximum capacity, an increase in the number of vehicles on the road leads to a decrease in speed and reduced flows. The difference between the average cost curve and the marginal cost curve represents the external marginal cost of congestion, i.e. the share of the congestion costs imposed by a road user on other motorists. The user chooses to use infrastructure according to the average cost of his trip, which results in "over-consumption", in that the costs borne by the user do not cover all the costs he generates. As a result, the optimum price, or congestion charge, corresponds to the difference between these two parameters.

1.5.2. *The problematic introduction of congestion charging*

From an economic standpoint, the need to charge for congestion would appear to be a foregone conclusion. It represents progress because it introduces the time related to congestion into the economic calculation made by agents. To justify the substitution of a price logic by free access, associated with tax-based financing, we can turn to Jules Dupuit[7], who is acknowledged as the first to advance the concept of specific charges for transport infrastructure which take account of contributory capacity. With regard to road infrastructure, a distinction needs to be drawn between infrastructure which has been in existence for a long time, and which is more or less amortized, and infrastructure which has not as yet been built or which must be financed even though it is relatively easy to identify its future users. This is the case, for example, for a bridge or a tunnel, both of which are civil works designed to meet a specific need at a specific location to improve traffic flow conditions. Jules Dupuit demonstrates that financing by the user is both possible (tolls) and more remunerative if the charge takes account of the contributory capacities of users, that is to say, if there is a certain degree of discrimination.

Congestion charging is a form of discrimination which consists in differentiating charges over time, according to the degree of congestion in the infrastructure, and therefore to move the charge along the distribution curve of time values. The person who is willing to pay more in order to travel in better flow conditions during peak hours, obtains greater utility than the person who prefers to pay less by changing mode or by changing the time at which he travels to off-peak hours. The temporal differentiation of pricing thus enables the community to benefit doubly:

– Firstly, infrastructure use is optimised by taking account of the differential utility to users. The price signal plays its role to the full by indicating relative scarcities and by selecting between those expressing demand;

– Secondly, it releases the financial resources to cover the cost of infrastructure.

Charging, differentiated according to levels of road congestion, can therefore help to steer both demand, by removing the users who create congestion and thereby reduce service quality, and supply, by giving priority to the construction of infrastructure whose costs can be covered by such charging. The three objectives that charging for public services generally seeks to meet (covering costs, steering demand and redistribution) are therefore concurrently taken into account by this type of charging system, which is why it has been applied for many years in the air transport sector and for high-speed services in the rail sector. Its use is also recommended in the road sector, particularly in urban areas, although such practices are far from widespread.

We must therefore ask ourselves why it is proving so difficult to apply such an effective solution in practice. Is it because public policies lag behind economic thinking? Or could it possibly be because the central role assigned to travel time, and therefore speed, in the above ways of thinking needs to be looked at anew?

Questions on the universal relevance of the price-time model primarily arise in urban areas when a critical look is taken at the objectives of charging.

– When the purpose of charging is to finance the construction of new urban roads, which are generally extremely expensive (tunnels, bridges, etc.), the toll must be set at a level which is generally unacceptable to the vast majority of users. Charging is then confronted with the problem of "three-sided incompatibility", which makes it impossible to have, at one and the same time, daily use of the infrastructure and high charging levels in the absence of any genuine alternative route. If the toll is to be socially acceptable, its level must be reduced. The outcome is insufficient revenue streams, requiring massive public transfers to road as a mode of transport, which is equivalent to subsidising road transport. The time savings and their economic implications thereby become a pretext for subsidies, which might be justified were it not for the conflict with other environmental (pollution, noise, etc.) or urban (urban sprawl, modal split, etc.) objectives.

– In the case of a "pure" congestion charge, designed to ensure a given level of traffic fluidity without building new infrastructure, it needs to be borne in mind that demand elasticity is relatively low during peak hours. Short of raising tolls to extremely high levels[8], the number of vehicles will still be high and the gain in travel speeds low. It should be noted that this low elasticity is more a sign of the captivity of motorists than a real willingness to pay. This was the main reason for the problems faced in introducing a congestion charge, as already pointed out by Baumol and Oates. On average, the congestion charge results in a net transfer to the community. The gain attributable to the time saving afforded by increased fluidity is more than offset by the cost of the charge, if the average value of time is taken as a basis. In other words, solely the small minority of individuals who have a very high time value benefit from a congestion charge, if the latter is introduced in a situation in which users have no other real alternative, in terms of route or time at which trips are made.

In a densely-populated urban area, efforts to use car speeds as a means of reducing the generalised cost therefore raise a number of questions. The pressure of demand for travel by car remains extremely high and any local improvement in fluidity results in an overall increase in traffic. Short of systematically oversizing the road network or imposing socially and politically unacceptable charges[9], it would be vain to adopt the vague and general objective of increasing fluidity, i.e. speed. What recent urban policies can teach us is that, on the contrary, we need to adopt a differentiated approach to the network. While on many routes it is appropriate to maintain a certain speed, the same is not true of the city centre and the roads that lead to the centre. In the first instance, it may be necessary to consider building new infrastructure. In the second, on the other hand, the aim of elected representatives, at present, would seem to be directed more towards reducing speeds, not only for safety reasons but also, and above all, in order to rehabilitate the urban environment. This reasoning might seem paradoxical and even anti-economic, since it consists in curbing the volume of traffic and in reducing the surface area of the road network in order to do so. However, we shall see in the second part of this paper that, under certain conditions, this objective is undoubtedly acceptable -- the problem being how to determine its area of relevance.

2. TRAVEL TIME: A CONSTANT IN ACTIVITY PROGRAMMES AND A DILEMMA FOR COLLECTIVE CHOICES

The objective of minimising generalised transport costs is a key factor in the understanding of individual and collective choices with regard to transport. This mechanism prompts users to give priority to the fastest modes of transport, those which will "save time". It should not be deduced from this, however, that users spend ever shorter periods of time in the transport system. The situation is quite the opposite since the time saved is, in a certain manner, reinvested in transport, as set out in Zahavi's hypothesis. Taking account of this hypothesis encourages us to look beyond the time spent on travelling and to take a closer look at the programmes of activity of individuals. It is quite plausible that the travel time gained in a typical trip is reinvested in additional distance (greater distance between home and the workplace) or in additional trips relating to new activities. In itself, this type of income effect is hardly surprising in economic analysis. However, to the extent that it has unwanted effects which challenge the sustainable nature of mobility, particularly in urban areas, the question arises as to whether the key objective of reducing the generalised cost of transport should not be replaced by the objective of increasing that very generalised cost, in some cases simply by reducing travel speeds!

2.1. Zahavi's hypothesis

In the 1970s, Y. Zahavi, an economist working at the World Bank, developed two hypotheses regarding the constancy of travel time and money budgets. His subsequent work on this double constancy led him to develop two different types of tool:

– A dataset which tended to confirm the regularity in relation to income of the time budget, on the one hand, and the money budget on the other;

– A traffic forecasting model (UMOT) in urban areas, based on data on the relative speeds of different modes and the trend in income.

The average travel time budget associated with Zahavi's hypothesis is approximately one hour. To be more precise, the result achieved by Zahavi consists in an approximation of the average travel time budget within a conurbation by means of a decreasing and very rapidly converging mathematical expression of average travel speed. The relationship between travel speed and average time budget is therefore represented by the following function:

$$T = b + \frac{a}{speed},$$

where T is the travel time per mobile person, and a and b are coefficients to be determined where b can be interpreted as the minimum time that an individual will assign to transport. The level of b will therefore be slightly below one hour of travel time.

For all the estimates made on the basis of different sample data, Zahavi obtained a very rapid convergence in the average travel time budget to one hour of travel time. In fact, as soon as average speed exceeds walking speed, the time budget appears to be convergent at slightly above one hour of daily travel.

Figure 10. **Travel time per mobile person and door-to-door speed**

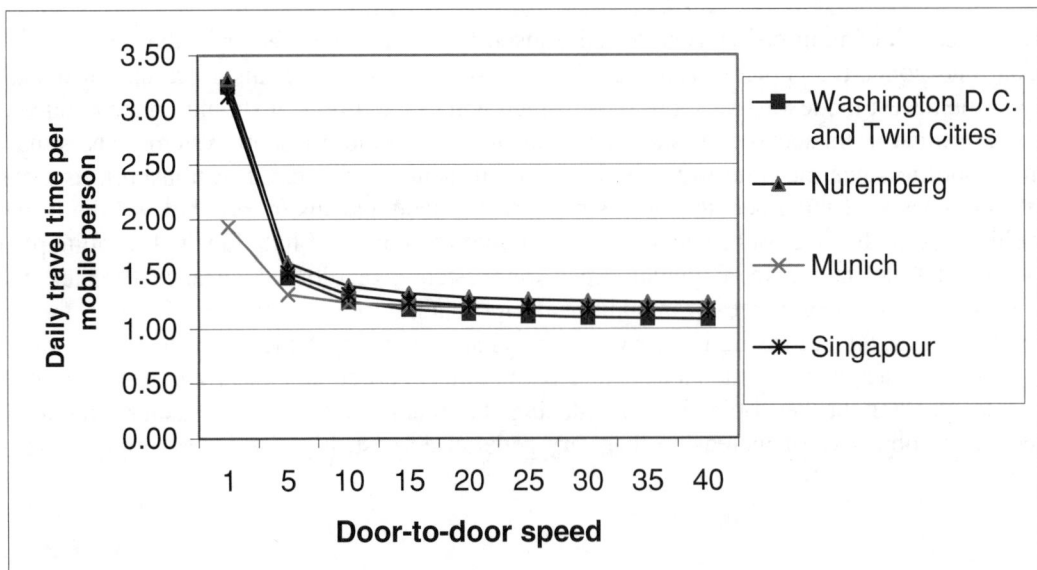

Round Table 127: Time and Transport – ISBN 92-821-2330-8 - © ECMT, 2005

As soon as the speed rises to 10 km/h, travel time budgets bunch together in a relatively narrow interval. The curves accept asymptotic values of b, which for these cities are relatively close ($b \in [1.03;1.18]$, in hours of travel time). The coefficient a indicates the speed at which the travel time budget will converge. The lower the value of a, the faster the time budget will decrease with speed ($a \in [2.01;2.18]$, except in Munich where $a = 0.77$). However, the level of the time budget depends upon the unit of observation used. For analyses based on mobile people, the time budget is slightly above one hour. However, it has been possible to "illustrate" the constancy of travel time budgets by means of observation units other than mobile people.

In an earlier paper[10], Zahavi focuses on the daily duration of car trips. Taking data from 18 different cities[11], Zahavi found that a critical level of car ownership appears to exist, above which the average duration of car trips are concentrated around the same average of 0.8 h per vehicle per day. The regularity of length of car trips would appear to be confirmed by data relating to vehicles. The duration of daily car trips in developed countries (cities with car-ownership levels above 10 per cent) are concentrated within a narrow time interval (0.70 h-0.88 h), i.e. 42 min-53 min.

The hypotheses and preliminary work by Zahavi have recently been confirmed by the results of work by A. Schafer (2000), who was able to benefit from improvements to transport surveys in the various locations studied. This allowed him to present constant daily travel time budgets in both time and space over a period extending from 1975 to 1997, despite the fact that the distances travelled vary substantially from one country to another and are rising over time.

Figure 11. **Travel time budgets, in hours per person per day**

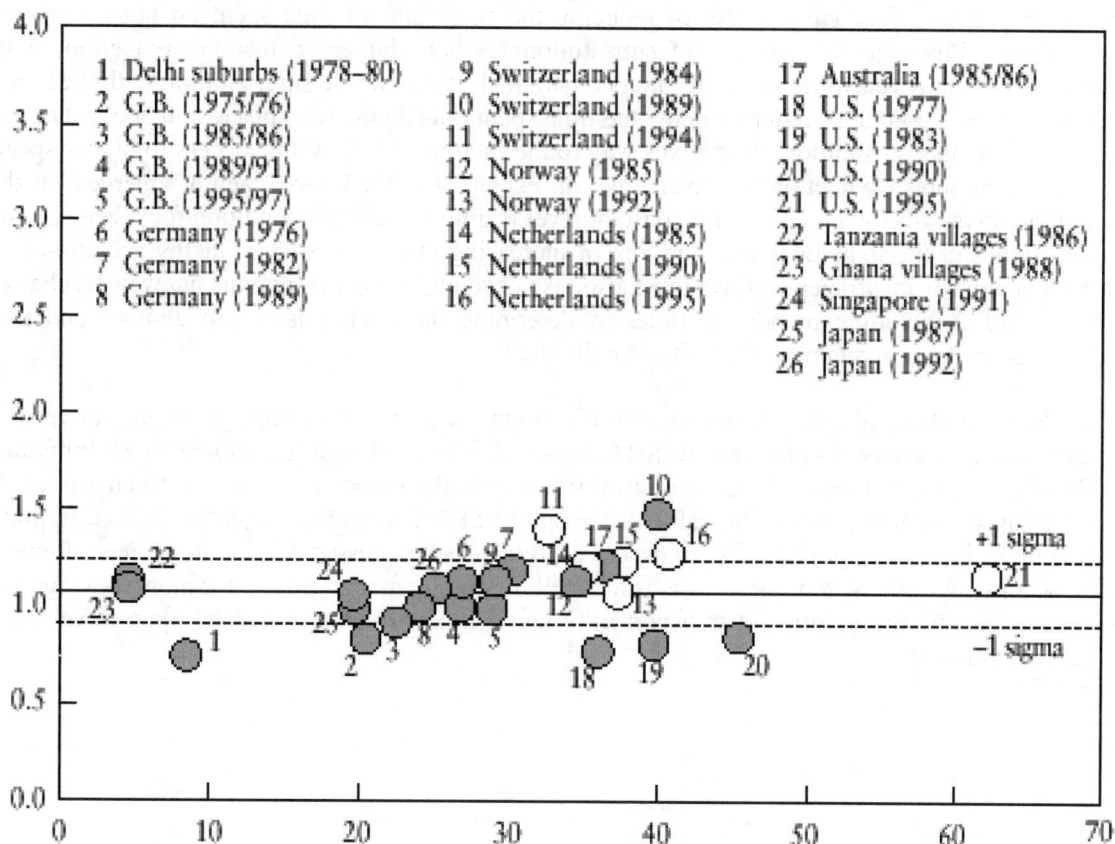

1 Delhi suburbs (1978–80)	9 Switzerland (1984)	17 Australia (1985/86)
2 G.B. (1975/76)	10 Switzerland (1989)	18 U.S. (1977)
3 G.B. (1985/86)	11 Switzerland (1994)	19 U.S. (1983)
4 G.B. (1989/91)	12 Norway (1985)	20 U.S. (1990)
5 G.B. (1995/97)	13 Norway (1992)	21 U.S. (1995)
6 Germany (1976)	14 Netherlands (1985)	22 Tanzania villages (1986)
7 Germany (1982)	15 Netherlands (1990)	23 Ghana villages (1988)
8 Germany (1989)	16 Netherlands (1995)	24 Singapore (1991)
		25 Japan (1987)
		26 Japan (1992)

A. Schafer conducted the same analysis for a wide variety of cities[12]. The figure below reveals an interval of forty minutes within which all the average travel time budgets of the cities studied are concentrated. The level of GDP in the countries surveyed has no significant impact on the level of the travel time budget and, for some cities for which a sufficiently long observation period is available, we can see an illustration of the continued regularity of the time budget, despite the economic growth of countries (Paris, Tokyo, Osaka).

2.2. Value of time and optimisation of activity programmes

It is important to note that, in the argument advanced by Zahavi, the constancy of travel time budgets must not be attributed to sociological or biological considerations. It is not a structural feature that we must accept by virtue, for example, of temporal-biological determinism. On the contrary, we are typically faced with a microeconomic process of optimisation. For proof of this, we need look no further than the fact that there is no convergence in cases where car-ownership levels are either low or zero. After the rise in travel speeds as a result of the use of cars and certain types of public transport, individuals modify their choices, firstly, in favour of higher average speeds that will reduce total travel times and, secondly, in favour of maintaining the same travel time, at either constant or increasing speed, while increasing the distances travelled, a simple but robust indicator of the activity opportunities afforded by trips.

The question of the relative stability of the travel time budget can be analysed, to some extent, as the development in the choice between working time and leisure time (in the broad sense of the term) in developed countries. According to the model proposed by G. Becker, the increase in real salaries encourages individuals, and women in particular, to opt for longer working hours that will make it possible to purchase products capable of reducing the constraint of time spent on housework and child-rearing. However, this process of substitution is short-sighted in that the reduction in the "constrained" time reveals the utility of "non-constrained" time (i.e. leisure, cultural activities, etc.) spent outside the workplace. After reducing the time spent outside the workplace, and above a certain level of income, the substitution effect gives way to the income effect. While access to higher speeds will initially prompt a reduction in travel time, an asymptotic trend subsequently emerges, in that maintaining travel time at a more or less constant level is quite simply the condition for diversification of the activities of the individual and therefore of increasing utility. However, taking account of the time budget alone is insufficient and we must also take account, as microeconomic analysis teaches us, of all resources, including money, in order to determine the central issues in choices and their implications for the programmes of activity of individuals.

In the most classical types of microeconomic model, two resources impinge on the universe of mobility choices, namely, income and available time. The level of mobility chosen by an individual will therefore be the outcome of the relationships between the component costs and benefits of the form of transport he uses. While the value of transport utility is a highly subjective notion, in that it depends upon individual preferences, transport costs can be expressed by a unit of measurement: money or time. And the size of these costs is dictated, firstly, by market prices, for monetary transport costs and, secondly, by the price of transport in terms of time, which will depend upon travel speed and the value of time.

Round Table 127: Time and Transport – ISBN 92-821-2330-8 - © ECMT, 2005

Figure 12. Average travel time budgets per person

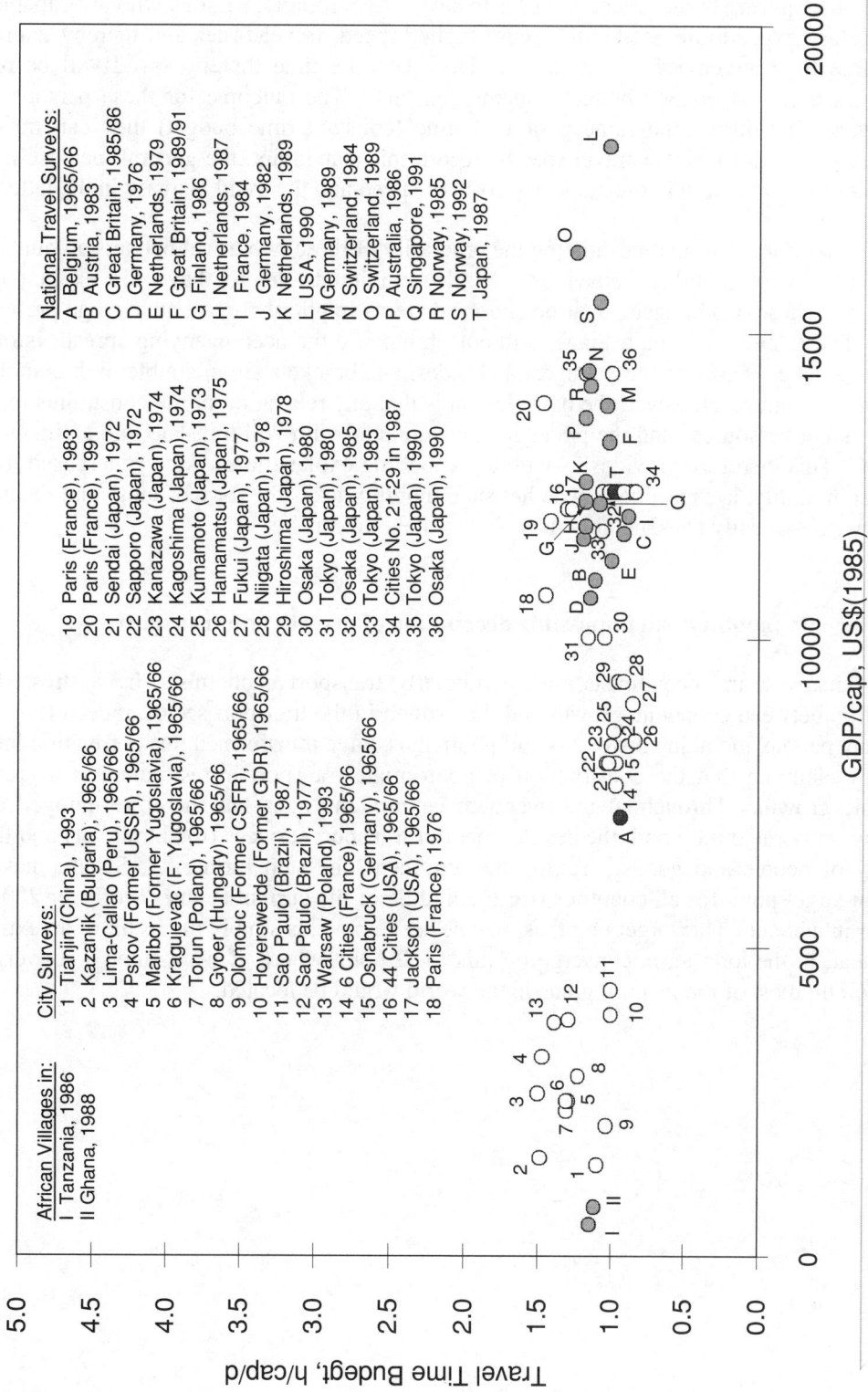

African Villages in:
I Tanzania, 1986
II Ghana, 1988

City Surveys:
1 Tianjin (China), 1993
2 Kazanlik (Bulgaria), 1965/66
3 Lima-Callao (Peru), 1965/66
4 Pskov (Former USSR), 1965/66
5 Maribor (Former Yugoslavia), 1965/66
6 Kragujevac (F. Yugoslavia), 1965/66
7 Torun (Poland), 1965/66
8 Gyoer (Hungary), 1965/66
9 Olomouc (Former CSFR), 1965/66
10 Hoyerswerde (Former GDR), 1965/66
11 Sao Paulo (Brazil), 1987
12 Sao Paulo (Brazil), 1977
13 Warsaw (Poland), 1993
14 6 Cities (France), 1965/66
15 Osnabruck (Germany), 1965/66
16 44 Cities (USA), 1965/66
17 Jackson (USA), 1965/66
18 Paris (France), 1976
19 Paris (France), 1983
20 Paris (France), 1991
21 Sendai (Japan), 1972
22 Sapporo (Japan), 1972
23 Kanazawa (Japan), 1974
24 Kagoshima (Japan), 1974
25 Kumamoto (Japan), 1973
26 Hamamatsu (Japan), 1975
27 Fukui (Japan), 1977
28 Niigata (Japan), 1978
29 Hiroshima (Japan), 1978
30 Osaka (Japan), 1980
31 Tokyo (Japan), 1980
32 Osaka (Japan), 1985
33 Tokyo (Japan), 1985
34 Cities No. 21-29 in 1987
35 Tokyo (Japan), 1990
36 Osaka (Japan), 1990

National Travel Surveys:
A Belgium, 1965/66
B Austria, 1983
C Great Britain, 1985/86
D Germany, 1976
E Netherlands, 1979
F Great Britain, 1989/91
G Finland, 1986
H Netherlands, 1987
I France, 1984
J Germany, 1982
K Netherlands, 1989
L USA, 1990
M Germany, 1989
N Switzerland, 1984
O Switzerland, 1989
P Australia, 1986
Q Singapore, 1991
R Norway, 1985
S Norway, 1992
T Japan, 1987

Travel Time Budegt, h/cap/d

GDP/cap, US$(1985)

Round Table 127: Time and Transport – ISBN 92-821-2330-8 - © ECMT, 2005

Representing mobility behaviour in this manner describes programmes of activity and individual travel according to five components: monetary costs, time costs, monetary resources and the value placed on travel utility. Any choice of programme of activity must be made with reference to these five components. There is, therefore, a certain type of trade-off between monetary costs and time costs when acquiring higher speed in order to save. Accordingly, persons who are capable of making the monetary expenditure needed to secure higher speed, travel faster and thereby gain in terms of travel time for a given level of mobility. However, the time thereby saved will be reinvested in transport, so that the money budget remains constant. The outcome for these persons is increased mobility, in that during the same period of time (constant time budget) they can travel a greater distance as a result of higher travel speeds. From this standpoint, the general increase in mobility is the result of the reduction in the monetary costs of speed and the trend increase in the latter.

It is therefore clear that postulating the regularity of these two travel budgets amounts to reducing the complexity of mobility behaviour. While individuals do not have the explicit objective of regularity in these two budgets, their behaviour patterns implicitly reveal an average preference of this nature. The reference to the average, without prejudging the accompanying spread, is of paramount importance here. Those in the youngest and oldest age brackets are undoubtedly less mobile than the working population. However, the basic lesson is that any relaxation of the constraints represented by the individual's resources and the prices he must contend with will translate into an increased level of mobility. This therefore provides us with a compelling explanation for the generalised trend towards increased mobility, irrespective of whether such mobility takes the form of long- and medium-distance inter-city trips or daily trips, notably those in urban environments.

2.3. Inter-city mobility: an impossible decoupling of growth and transport

For many years, economists, and particularly transport economists, have stressed the strong correlation between economic growth and the growth in the transport sector, indeed to such an extent that many people, including historians and politicians, have transformed this correlation into cause and effect by claiming that the construction of appropriate transport infrastructure is a prerequisite for economic growth. Throughout the twentieth century this idea has therefore enjoyed considerable currency, as borne out by both the development of transport services (road, rail, air) and the increased mobility of people and goods. Taking the argument one step further, A. Schafer has proposed a common target point for all countries (for the middle of the current century?) of over 200 000 km per year per inhabitant. This target point, as may be seen above, is simply the result of an extrapolation of trends that, in the long term, converge on this focal point where, if we assume a convergence in per capital GDP, most of the major regions in the world would be located.

Figure 13. **Correlation between growth and individual mobility: trends**
Total mobility in passenger-km per year
(Statistics 1960 - 1990; Trends 1960 - 2050)

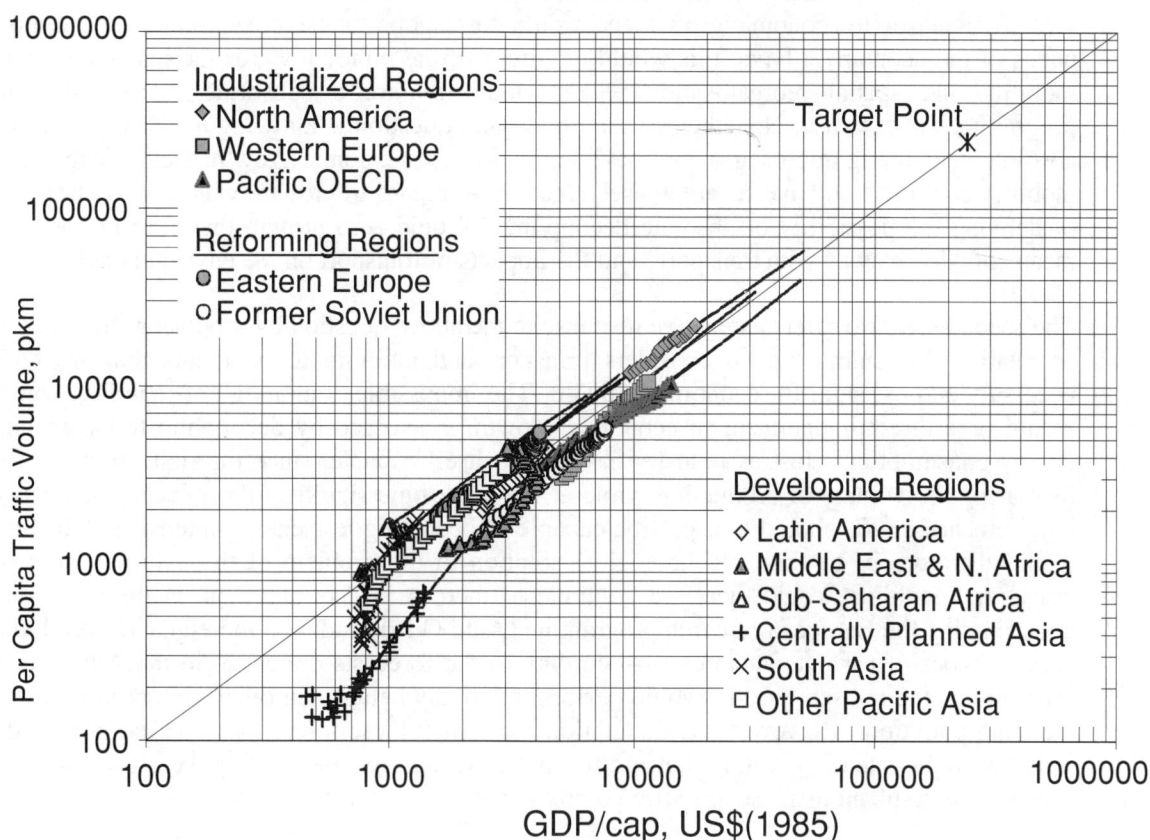

Source: Schafer and Victor (2000); economic growth rates based on IPCC IS92a/e scenario.

This is clearly a scenario that gives serious cause for concern in view of the impacts that such a dramatic explosion in mobility will have on the environment and/or land use: 200 000 km per year, of which almost a third by high-speed modes of transport (high-speed train and air transport), given that the modes of transport experiencing the strongest growth are also the fastest and the most intensive in terms of energy consumption and therefore those which generate the most pollution (see Figure 1). Let us recall that, excluding walking, per capita mobility since 1800 in the United States has grown at an annual rate of 4.6 per cent, and by 2.7 per cent if pedestrian traffic is included. By way of comparison, the mobility of the French has risen at an annual rate of 4 per cent since 1800[13].

These concerns have given rise to the idea that the link between economic growth and mobility should be severed, a proposal in direct opposition to the argument advanced by A. Schafer. For the time being, the concept of decoupling conveys vague and implicit connotations. What would be the ultimate objective of such a process? Would it be what is termed "absolute decoupling" or would it simply mean severing the link between mobility and energy consumption, or what is termed "relative decoupling"?

- The first scenario would involve an absolute constraint on per capita mobility, possibly by establishing quotas for per capita mobility, in that the main reason to reduce or stabilize mobility is to curb emissions of CO_2 and other pollutants. Absolute decoupling does not mean zero mobility, but rather the stabilization of mobility in absolute terms and, what is more, even during periods of economic growth. If decoupling is viewed as a dynamic concept (elasticity of annual growth in GDP and growth in traffic levels), this means that mobility must be virtually inelastic to economic growth and would correspond to a stabilization of the transport needs of the economy. To do this would require a strong policy towards the management of mobility and spatial reorganisation which would require a substitute for mobility in programmes of activity. Nonetheless, the underlying question is knowing to what extent this absolute decoupling approach would have an impact on economic growth itself. Stabilizing mobility could indeed have severe consequences with regard to growth, which is another way of denying the possibility of absolute decoupling. Would zero growth therefore be the sole means of "doing away with transport" and the impacts of transport on the environment?

- The second scenario offers a negative response to the above question by proposing the solution of relative decoupling, which considers transport and mobility to be inputs that are both necessary and essential for economic growth. The impossible "rationing" of mobility, which would be a direct impediment to economic growth, is avoided by decoupling transport and energy consumption. Just as industry has proved itself capable, since the first oil shock, of increasing the volume of production while at the same time significantly reducing its energy requirements, so too should transport be capable of becoming less energy-intensive. The aim would therefore be to find a technological solution to the problem of the external costs of transport. Mobility would therefore continue to increase. This argument is the same one underpinning the European research programme (Auto Oil II) and, at a more general level, the current hopes of car manufacturers. According to the forecasts drawn up in the Auto Oil II programme, the transport sector would perform relatively better than other sectors in terms of reducing pollution. However, the assumptions on which these forecasts are based are highly sensitive to the rate of growth in demand for mobility. Relative decoupling therefore leaves to one side the resultant increase in traffic volumes, congestion and spatial impacts.

The technological solution would, at first sight, seem appealing since it would simply be a prolongation of current ways of living and thinking. However, it contains the seeds of its own limitations within itself. Such a solution would generate mobility, given that, because the cost of technology constantly declines as a result of productivity gains, we would see the continued operation of the mechanism whereby additional spending power is devoted to the purchase of a faster form of mobility. In addition, the time savings afforded by the optimisation of trips could work to the same effect by pushing back the limits to infrastructure congestion and saturation. Ultimately, the impact that economic growth has on programmes of activity is to achieve a spatial transfer of the problems posed by the regularity of travel time budgets. If, far from being solely a variable which needs to be minimised, travel time is also a constant, then increased travel speeds will lead to increased spatial occupation by individuals. The growth in the absolute value of the money budget for mobility will broaden the scope and increase the number of trips. Through rapid development of high-speed modes of transport and the inexorable growth in leisure mobility, medium- and long-distance mobility clearly has great prospects for the future, as the hypothesis of "double constancy" would appear to suggest to us.

2.4. Daily mobility: travel speed and activity programmes

While the prospect of future growth in inter-city mobility is not a major cause for concern, the same cannot be said of urban mobility. Such mobility is growing in accordance with the same basic principles as medium- and long-distance mobility. As O. Morellet and P. Maréchal have shown, it would be perfectly feasible to include both of these two forms of mobility in a single analytical model (cf. the MATISSE model), capable of efficiently predicting transport demand by classifying demand by major types of trip, primarily grouped according to their price sensitivity and travel times. Overall, wherever speed gains are technically feasible and financially accessible, modes with the lowest generalised costs attract most support from the population. This is notably the case for car use in urban areas. Even though the analytical bases are the same, the implications, notably in spatial terms, of the increase in urban mobility pose specific problems.

In the first analysis, we should welcome the growing role played by the car in the organisation of our mobility, in that it is the choice which enables generalised costs to be minimised. This positive assessment could even be enhanced by introducing the benefits of a trip. It goes without saying that reducing the average generalised cost can also lead to the emergence of new travel opportunities. At a lower cost, certain trips will become attractive, and it is at this point that we come up against the issue of the sustainability of car-based mobility. How can we avoid car-based mobility leading to increased mobility in the form of trips over longer distances, either for leisure pursuits, work or, primarily, choice of place of residence?

To illustrate the latter point, let us take the case of journey-to-work trips in France made by persons who work in a different commune to the commune where they reside. By comparing data from the 1990 and 1999 surveys, it appears that the total distance travelled by intercommunal migrants rose from 165 million to 211 million kilometres a day, an increase of almost 28 per cent in less than ten years. This growth is partly attributable to the longer average distances travelled daily (approximately 7 per cent), but is mainly due to the number of persons making trips (almost 20 per cent). Since there are two movements which are cumulative, the main problems regarding the sustainability of daily mobility are clearly to be found here.

Note that the trend in distances travelled and the number of alternating migrants vary substantially according to place of residence. As may be seen in the table above, the rates of variation differ widely from one area to another, resulting in structural effects.

-   A first type of structural effect derives from the fact that residents in urban poles (cities with a centre and suburbs) on average travel shorter distances than others. However, since the rate of increase in the number of intercommunal migrants is much higher in peripheral and rural areas, the total distance travelled is rising at a faster rate in this second category of location (+35.8 per cent as opposed to +20.6 per cent in urban poles), which now accounts for a slightly higher total distance travelled.

-   The same phenomenon can be seen within urban poles. In a rather curious development, the average distance travelled is rising faster than the average distance travelled by residents in town centres, on the one hand, and suburban residents on the other. This may be explained by the simple reason that the first category of residents, who paradoxically make the longest trips, is growing faster than the second. It is therefore clear that control over urban mobility, and private-car mobility in particular, must take account of structural phenomena.

Table 3. **Intercommunal journey-to-work trips in France (1990-99)**

	Total daily distance (thou. kms)	Rate of growth 99/90	Number of daily migrants (thousands)	Rate of growth 99/90	Average daily distance (km)	Rate of growth 99/90
City centres	36 982	+28.0%	1 988	+ 21.8 %	18.6	+ 5.0 %
Suburbs	68 887	+18.2%	5 939	+10.1%	11.6	+ 7.4 %
Total urban centres	105 869	+ 20.6%	7 927	+ 12.7%	13.3	+ 8.1 %
Of which: *urban area of Paris*	*35 555*	*+11.8 %*	*2 914*	*+ 8.1 %*	*12.2*	*+ 3.4%*
Outskirts	52 003	+34.0%	3 133	+ 29.3 %	16.6	+ 3.8 %
Of which : *urban area of Paris*	*12 828*	*+ 23.8 %*	*539*	*+ 24.5 %*	*23.8*	*+ 1.3%*
Multi-centre urban areas	15 382	+39.0 %	855	+ 31.3%	18.0	+ 5.9 %
Rural areas	39 377	+36.7%	2 128	+ 33.1 %	18.5	+ 2.8 %
Total excl. urban centres	106 762	+35.8 %	6 116	+ 30.9 %	17.5	+ 3.7 %

Accordingly, current patterns of urban development, which are more extensive and often multipolar, require the population to make relatively long trips on extremely varied routes. Even if the public transport system is efficient, it remains insufficient to cope with demand for mobility, which can only be met by car use. And the more the share accounted for by car use increases, the more it encourages households and firms to move to peripheral locations, there commensurately increasing dependence on car use. Indeed, this explains the relatively high average daily trip distance for the inhabitants of city centres.

Urban structure and mobility are therefore closely linked.

– Firstly, the urban structure provides the space in which all the possible activities pursued by individuals are distributed. The locations of these different activities in relation to the locations of individuals induce certain patterns of mobility. As Zahavi points out, differences in the distribution of employment zones and residential areas may provide an indication of the minimum level of mobility that will obtain. Concentrations of activities, relating either to work or leisure, will generate a large share of trips, and the locations of such concentrations will have an impact on patterns of mobility. Peripheral locations, for example, will lie beyond the reach of public transport services and will therefore be prime sites for generating trips by car.

Round Table 127: Time and Transport – ISBN 92-821-2330-8 - © ECMT, 2005

– Secondly, the urban structure is also where transport will take place and will therefore, in large part, dictate the conditions under which trips are made. In many cases, urban density is synonymous with reduced travel speeds. M. Wiel (1999) accordingly speaks in terms of co-generation between the town and mobility. In his view, urban sprawl can be explained at least in part by the development in increased travel speeds. One of the outcomes of increased travel speeds is an increase in the distances travelled, without any changes in the travel time budget. Urban sprawl, as a response to the phenomenon of property rents, would therefore be the outcome of individuals' taste for increased living space in their homes by extending the limits to their area of mobility.

According to this logic, the way to put a brake on urban sprawl would therefore be to impose speed restrictions, since if speed is responsible for the extension of cities then it should also be capable of slowing down, if not reversing, urban sprawl. Lower speeds should reduce the spatial access of individuals and, as a result, encourage them to modify their choice of place of residence, to the benefit of centres of activity. Such a decision would be expected to restore the density of cities and city centres (both former and new centres)[14]. While solutions are rarely advocated so starkly (i.e. reduce the number of daily trips), it is clear that this type of reasoning has inspired many of the strategic analyses conducted both by experts[15] and by local elected representatives[16]. Both parties based their analyses on the impossible headlong rush towards "ever more" mobility in urban areas.

Reducing speeds has only a limited area of relevance, however. It is perfectly feasible to envisage reducing traffic speeds in city centres, provided that the performance level of alternative modes is sufficient to meet mobility needs. Instruments such as park-and-drive and public transport lines of sufficient capacity can make speed restrictions on car traffic in city centres an operational reality. If public transport can capture a share of the traffic, the increase in congestion in densely populated areas in the centre can be halted. In some ways, restrictions in city speeds can enable the speed of public transport systems to retain their comparative advantage and to limit the return of car traffic.

Introducing this type of measure in peripheral areas is more problematic. On urban ringroads, the supply of many to-and-from trips cannot reasonably be ensured by public transport. Reducing speeds on roads into the centre could encourage a modal shift to public transport for trips towards the centre. However, it could also severely disrupt trips from one peripheral area to another, where public transport cannot readily compete due to the conflict between the level of service required (strong demand for flexibility, frequency, etc.) and the level of patronage which will ensure minimum economic viability.

All of these considerations would clearly seem to show that the question of car speeds in urban areas leads us directly back to the social choices made with regard to the value of time. By the same token that a social value exists for the discounting rate and a social value for time in economic calculations, it all seems as though urban policies currently aimed at promoting public transport were making an adverse selection of time values. Where economic analysis of congestion suggests that, in order to increase fluidity, charges should be introduced in an attempt to eliminate users whose time values are too low, urban policies are substituting a different approach. Without explicitly seeking to drive away users with high time values from city centres, these policies propose another form of arbitration.

What, precisely, are the reasons for the trips made by users driving in the city centre? Whether the trip is from (or towards) his home, from (or towards) his workplace or from (or towards) a shopping or recreational area, the user must be made aware of the fact that the amenities he is seeking in that area are incompatible with the speed of trips by car. At the risk of caricature, it is as though the elected representatives of city centres were increasingly tempted by a social vision of time value of the type that prevails in "Disneyland[17]" or the centre of areas that attract large numbers of tourists. The key signal sent to users of these spaces is that a relative degree of slowness is the price that has to be paid to use the collective good constituted by the theme park. The development of tramways, a relatively slow mode of transport compared with underground railways, is an example of this choice which can be seen today in a very large number of French and also European cities (Barcelona, Geneva, etc.). Abandoning the headlong rush towards higher speeds, what users are offered is a certain quality of urban life which is based on the assumption that trips will be relatively slow. Users (both households and firms) are therefore called upon to reorganise their programme of activity by modifying their route, by making trips at earlier or later times, by changing their place of residence or even by changing their workplace.

The interest in this new configuration, which has already been well-established for many years in cities such as Berne (Switzerland) or Freiburg and Karlsruhe (Germany), is that the decision to make such choices has not ended up stifling the life in city centres. On the contrary, property prices in central areas have risen and their attractiveness as both business and residential locations has remained unabated. The local elected representatives have therefore made a coherent economic choice, aimed at enhancing the value of the public assets they are in charge of managing. The value they have most enhanced is not time but the property assets located within the urban area and they have done this in terms of property values. Given a choice between time and space, they have opted for the latter and this priority takes precedence over the time values implicitly required by users.

2.5. Conclusion: Towards charging for trips at the generalised cost?

We must clearly repeat here that the decision at the local level to reduce car speeds and to place a low collective value on time is not an anti-economic decision, even though it might initially appear to be so. At any rate, it is not anti-economic provided that the area subject to such restrictions is not too large. Just as time values are low in the very centre of Disneyland but high in relation to the access roads which lead there[18], so too the low speed of trips within the city centre are accepted all the more readily in that there are genuine possibilities for transit traffic or suburb to suburb traffic to avoid the centre. It is for this reason that those elected representatives who wish to impose severe constraints on car mobility in the city centre, militate at the same time for the construction of ringroads and motorway bypasses, even if they are extremely costly. There is no inconsistency in such an attitude in that their constraint on speeds and time values does not represent a universal position of principle. At most, it is a contingent choice related to management of property assets in a specific spatial location.

Even though, when presented in this way, this choice is perfectly rational, it is by no means certain that politicians have fully grasped all the implications of such a choice, in that, if we now place ourselves at the level of the conurbation and not the city centre, the model of the spatial segmentation of time values leads to neither a reduction in mobility nor to a stabilization of transport infrastructure needs. While many local transport policies claim to want to substitute investment in public transport for investment relating to private car use, the situation we find ourselves facing is one of complementarity, where we need to increase both components. However, in urban and outlying areas, investment in transport infrastructure comes at a very heavy price. The increased demand for mobility will therefore lead to a heavier cost burden on the community, which will have to be passed on, in one way or another, to users.

Seen in this perspective, charging from trips made in urban areas assumes an entirely new dimension. While the economic rationale for congestion charging is still based on the time savings afforded by a higher degree of road pricing (time versus money), we are gradually moving towards a situation in which, in some cases, it is perfectly legitimate for the two components of the generalised cost of a trip to move in the same direction, namely, towards an increase in the total cost. While such a possibility seems unacceptable from a purely individual point of view, it makes sense if it is part of an overall urban plan, as shown by the examples of large urban areas (London, Oslo, Tronheim, etc.) which have experimented with urban tolls based on the principle of area charges rather than the concept of explicit congestion charging.

As in these conurbations, urban transport policies will, in future, work on several mechanisms at the same time: a differential, although trend, decrease in the average speed of cars in urban areas; highway building restricted to bypass roads; introduction of charges on trips made by car; development of public transport, including intermodal configurations which will help steer the residents of peripheral areas to a small number of major corridors. All of these measures will obviously come at a cost to public finances. Increasing the cost of mobility is therefore a necessity, given that investment is required, not to mention subsidies for public transport. However, this increase cannot, under any circumstances, be seen as a cornucopia that will allow local authorities to reduce other taxes, for example. By the same token, this will not make any fundamental changes to urban forms and the trend towards development on the periphery of towns and cities; in such cases, a supply of efficient public transport services will add to the severity of the problem. No radical changes should therefore be made to current trends, nor should car mobility be dealt a death blow; instead, a coherent set of signals needs to be sent to users. The collective urban project requires that both components of the generalised cost of trips by car -- namely, price and duration -- must be increased, at least at the local level. This can be achieved gradually and in some respects has already been initiated. Surprising at it might seem, the issue at stake consists no more or less than in applying a simple economic principle: increasing the cost in response to scarcity, that environmental considerations could make even more severe.

NOTES

1. We shall dismiss here the argument, advanced by Ivan Illich, to the effect that time attributable to speed would be completely absorbed by the additional work time needed to purchase that speed. The lower unit cost of travel in relation to the average wage is one aspect of economic growth that is not a zero-sum game.

2. A comparison with different European values is available in the Boiteux 2 report.

3. The breakdown of the average value for all trip reasons was as follows: professional trips 10%; journey-to-work trips 35%; trips for other reasons 55%.

4. Harvey proposed the following type of formula: discounting rate = $a0*b/(b+t)$, where $a0$ is the discounting rate for the year in which the service first starts up, b a constant and t the time after entry into service. Heal introduced logarithmic discounting, which was then formulated by Overton and MacFadyen.

5. See Cline, W., 1992 and 1999.

6. To give an order of magnitude, the maximum flow rate for an urban clearway is 1 800 vehicles per hour and per carriageway at 55 km/h (Hau, 1998).

7. For a detailed presentation of the argument advanced by J. Dupuit, see M. Allais (1989).

8. This is the case of the congestion charge recently introduced in central London (five pounds sterling), whose residents pay solely 10 % of the charge.

9. In this respect, the toll introduced in Singapore is more of a deterrent than a model.

10. Zahavi, Y. (1973), The TT-relationship: a unified approach to transportation planning, *Traffic Engineering and Control*, pp. 205-212.

11. Athens, Baltimore, Baton Rouge, Bombay, Brisbane, Chicago, Columbia, Copenhagen, Kansas City, Kingston, Knoxville, London, Meridian, Pulaski, Saint Louis, Tel Aviv, Tucson, West Midlands.

12. African villages (Riverson and Carapetis, 1991), 44 cities (Szalai *et al.*, 1972; Katiyar and Ohta, 1993; EIDF, 1994; Malasek, 1995; and Metrõ, 1989) and national data (Kloas *et al.*, 1993; Vliet, 1994; UK Department of Transport, Federal Highway Administration, 1992; Stab für Gesamtverkehrsfragen, 1986; Dienst für Gesaktverhekhrsfragen, 1992; Orfeuil and Salomon, 1993; Vibe, 1993; Federal Office of Road Safety, 1998; Olszewski *et al.*, 1994).

13. A. Gruebler (1990).

14.   This type of reasoning illustrates the new awareness brought about by the famous curve in which Newman and Kenworthy demonstrate the existence of an inverse relationship between density and per capita energy consumption.

15.   See Bieber, Massot and Orfeuil (1993), Kaufmann (2000) and also DRAST, Groupe de Batz (2002).

16.   See the Revue 2001 Plus, DRAST (No. 58, February 2002).

17.   It is interesting to note that each Disneyland is designed like a town.

18.   Eurodisney, in Marne-la-Vallée (Ile de France), has a direct link to the TGV and RER (high-speed regional railway in the Paris region) and Roissy Charles de Gaulle airport is not far away.

BIBLIOGRAPHY

Abraham, C. (1961), La répartition du trafic entre itinéraires concurrents : réflexions sur le comportement des usagers, application au calcul des péages, *Revue générale des routes et des Aérodromes*, No. 357, October, 39 pp.

Allais, M. (1981), *Théorie générale des surplus*, Presses Universitaires de Grenoble, 716 pp.

Arnott, R., A. De Palma and R. Lindsey (1998), Recent developments in the bottleneck model, in: Button, K. and E. Verhoef (1998), *Road Pricing, Traffic Congestion and the Environment*, Aldershot: Elgar.

Arrow, K. and M. Kurz (1970), *Public investment, the rate of return, and optimal fiscal policy*, Hopkins Press, Baltimore, 218 pp.

Arrow, K., R. Solow, P. Portney, E. Leamer, R. Radner and H. Schuman (1995), Report of the National Oceanic and Atmospheric Administration (NOOA) Panel of Contingent Valuation, *Federal Register*, No. 58, pp. 4601-4614.

Ausubel, J.H., C. Marchetti and P.S. Meyer (1998), Toward green mobility: the evolution of transport, *European Review*, Vol. 6, No. 2, pp. 137-156.

Baumol, W.J. and W.E. Oates (1988), *The theory of environmental policy*, Cambridge University Press.

Baumstark, L. and A. Bonnafous (1998), La relecture théorique de Jules Dupuit par Maurice Allais face à la question du service public, Paper given at the conference on "La tradition économique française -- 1848-1939", Lyon, 2-3 October, 15 pp.

Becker, G. (1965), Time and Household production: a theory of the allocation of time, *Economic Journal* 75, September, pp. 493-517.

Beesley, M.E. (1965), The value of time spent in travelling: some new evidence, *Economica,* Vol. 45, May, pp. 174-185.

Bieber, A., M.H. Massot and J.P. Orfeuil (1993), Prospective de la mobilité urbaine, in: Bonnafous, A., F. Plassard and B. Vulin (éds.) : *Circuler demain*, La Tour d'Aigues, DATAR, Ed. de l'Aube, coll. Monde en cours.

Blayac, T. and A. Causse (2002), Value of travel time, *Transportation Research Part B*, pp. 367-389.

Bonnafous, A. (1999), Infrastructures publiques et financement privé : le paradoxe de la rentabilité financière, *Revue d'Economie Financière*, No. 51, pp. 157-166.

Bourdaire, J.M. (2000), *Le lien entre consommation d'énergie et développement économique*, World Energy Council, April.

CEC (1998), *Towards Fair and Efficient Pricing in Transport*, Brussels.

Charpin, J.M., B. Dessus and R. Pellat (2000), Report prepared for the Prime Minister, *Etude économique prospective de la filière nucléaire, Annexe 8: Le choix du taux d'actualisation*, pp. 261-269.

Commissariat Général du Plan (1994); Report by the Group chaired by Marcel Boiteux, *Transports : pour un meilleur choix des investissements*, 131 pp.

Commissariat Général du Plan (2001); Report by the Group chaired by Marcel Boiteux, *Transports : choix des investissements et coût des nuisances*, 325 pp.

Crozet, Y. and G. Marlot (2001), Péage urbain et ville durable : figures de la tarification et avatars de la raison économique, *Les Cahiers Scientifiques du Transport*, No. 40, pp. 79-113, Editions de l'AFITL.

De Palma, A. and C. Fontan (2001), Choix modal et valeurs du temps en Ile de France, *Recherche, Transports, Sécurité*, No. 71, April-June, pp. 24-47.

Dijst, M. (2001), *ICTS and accessibility: an action space perspective on the impact of new information and communication technologies*, Paper presented at the 6th NECTAR Conference, 16-18 May 2001, Helsinki, Finland.

DRAST (2002), *Mobilité urbaine: cinq scénarios pour un débat*, "Groupe de Batz", 66 pp.

DRAST (2002), *Les politiques de déplacement urbain en quête d'innovations (Genève, Naples, Münich, Stuttgart, Lyon)*, in: *Revue 2001 Plus, Veille internationale*, No. 58, February, 52 pp.

ECMT, *Round Table 81, Private and Public Investment in Transport*, Paris, 1990, 119 pp.

Else, P. (1981), The theory of optimum congestion taxes, *Journal of Transport Economics and Policy*, Vol. 15, No. 3.

Evans, A. (1992), Road congestion pricing: when is it a good policy?, *Journal of Transport Economics and Policy*, Vol. 26, No. 3.

Giuliano, G. (1992), An assessment of the political acceptability of congestion pricing, *Transportation*, Vol. 19, No. 4.

Gruebler, A. (1990), The rise and fall of infrastructure: dynamics of evolution and technological change in transport, *Physica*, Heidelberg.

Harvey, C.M. (1994), The Reasonableness of Non-Constant Discounting, *Journal of Economic Public Theory*, Vol. 53, pp. 31-51.

Hau, T. (1992), *Economic fundamentals of road pricing: a diagrammatic analysis*, WPS 1070, The World Bank, Washington DC, pp. 1-96.

Hau, T. (1998), Congestion pricing and road investment, in: Button, K. and E. Verhoef (1998), *Road Pricing, Traffic Congestion and the Environment*, Aldershot: Elgar.

Heal, G.M. (1993), *The optimal use of exhaustible resources*, Handbook of Natural Resources and Energy Economics, Vol. III.

Henderson, J. (1974), Road congestion: a reconsideration of pricing theory, *Journal of Urban Economics*, 1, pp. 346-365.

Hensher, D.A. (2001), The valuation of commuter travel time savings for car, *Transportation*, pp. 101-118.

Hensher, D.A. (2001), Measurement of the valuation of travel time savings, *Journal of Transport Economics*, pp. 71-98.

Hotelling, H. (1931), The economics of exhaustible resources, *Journal of Political Economy*, No. 39, pp. 137-175.

Hotelling, H. (1938), The General Welfare in Relation to Problems of Taxation and of Railway and Utility Rates, *Econometrica*, No. 6, pp. 242-269.

Illich, I. (1975), *Energie et Equité*, Editions du Seuil.

Institute for Transport Studies, University of Leeds (October 2000), *Separating the Intensity of Transport from Economic Growth*, Report on the Workshop, "La Sapienza" University, Rome.

Kaufmann, V. (2000), *Mobilité quotidienne et dynamiques urbaines*, Presses Polytechniques et Universitaires Romandes, Lausanne.

Lave, C. (1994), The demand curve under road pricing and the problem of political feasibility, *Transportation Research*, Vol. 28A, No. 2.

Lenntorp, B. (1976), Paths in space-time environment: a time geographic study of possibilities of individuals, Department of Geography, The Royal University of Lund, *Lund Studies in Geography, Series B, Human Geography*, No. 44.

Lesourne, J. (1972), *Le calcul économique, théorie et application*, Editions du Seuil, Paris, 459 pp.

Lind, R.C. (1990), Reassessing the Government's Discount Rate Policy in light of new Theory and Data in World Economy with a high degree of Capital mobility, *Journal of Environmental Economics and Management*, No. 18, pp. 8-28.

Masse, P. (1988), Public Utility Pricing, *New Palgrave Dictionary of Economics*, MacMillan, London.

Masse, P. (1946), *Les réserves et la régulation de l'avenir*, Herman, Paris.

Mohring, H. and M. Harwitz (1962), *Highway Benefits: An analytical framework*, Northwestern University Press, Evanston.

Morellet, O. and P. Marechal (2001), Demande de transport de personnes : une théorie unifiée de l'urbain à l'interurbain, *Recherche, Transports, Sécurité*, No. 71, April-June, pp. 49-99.

Nowlan, D. (1993), Optimal pricing of urban trips with budget restrictions and distributional concerns, *Journal of Transport Economics and Policy*, Vol. 27, No. 3.

Orfeuil, J.P. (1999), *Evolution des mobilités locales et interface avec les stratégies de localisation*, PUCA.

Orfeuil, J.P. (2000), *L'évolution de la mobilité quotidienne*, Les collections de l'INRETS, No. 37.

Ramsey, F. (1928), A mathematical theory of saving, *Economic Journal*, No. 38, pp. 543-559.

Schafer, A. and D.G. Victor (2000), The Future mobility of the world population, *Transportation Research*, A 34, pp. 171-205.

Schafer, A. (2000), Regularities in travel demand: An international perspective, *Journal of Transportation and Statistics*, December.

Segonne, C. (2001), Choix d'itinéraires et péage urbain: le cas du tunnel Prado-Carénage à Marseille, *Recherche, Transports, Sécurité*, No.°71, April-June, pp. 2-23.

Talbot, J. (2001), Les déplacements domicile-travail, de plus en plus d'actifs travaillent loin de chez eux, *INSEE Première*, No. 767, April.

Verhoef, E. (1994), External effects and social costs of road transport, *Transportation Research*, Vol. 28A, No. 4.

Verhoef, E., P. Nijkamp and P. Rietveld (1995), Second-best regulation of road transport externalities, *Journal of Transport Economics and Policy*, Vol. 29, No. 2.

Walters, A.A. (1961), The theory and measurement of private and social cost of highway congestion, *Econometrica*, Vol. 29, No. 4, pp. 676-699.

Walters, A.A. (1988), Congestion, *New Palgrave Dictionary of Economics*, MacMillan, London.

Wiel, M. (1999), *La transition urbaine, ou le passage de la ville pédestre à la ville motorisée*, Edition Architecture et Recherches/Mardaga, 149 pp.

Wiel, M. (2002), *Ville et automobile*, Edition Descartes & Cie, Paris, 140 pp.

Zahavi, Y. and A. Talvitie (1980), Regularities in Travel Time and Money, *Transportation Research Record 750*, pp. 13-19.

Zahavi, Y. (1979), *The "UMOT" Project*, Report prepared for the US Department of Transportation and the Ministry of Transport of the Federal Republic of Germany.

THE VALUE OF FREIGHT TRANSPORT TIME: A LOGISTICS PERSPECTIVE – STATE OF THE ART AND RESEARCH CHALLENGES

Lorant A. TAVASSZY
TNO Inro
Delft
Netherlands

Nils BRUZELIUS
Nils Bruzelius AB
Lund
Sweden

THE VALUE OF FREIGHT TRANSPORT TIME: A LOGISTICS PERSPECTIVE STATE OF THE ART AND RESEARCH CHALLENGES

SUMMARY

Delft/Lund, July 2003

ACKNOWLEDGEMENTS

The authors gratefully acknowledge suggestions and comments on earlier drafts of the paper by Cees J. Ruijgrok of TNO Inro, The Netherlands.

1. INTRODUCTION

1.1. Freight VOT in the context of transport policy

The main application of the concept of value of time (VOT) is in impact assessments of transport policies, related to the valuation of reductions in door-to-door transport time. As is well-known from infrastructure Cost-Benefit Analysis (CBA) studies, short time savings can add up (over the years and over individual trips) to billions of benefits. In order to have a comprehensive account of the full value of time, i.e. not to miss any of the costs or benefits of time changes, we must be clear about the scope of our measurements: whose bookkeeping are we looking at; is it a social welfare value, a firm level value, a value for a supply chain? Clearly, the value of a time saving will be different for a carrier than for society as a whole. In this paper, we focus on the VOT within the context of the social cost-benefit analysis of transport projects and policies, i.e. we are looking for a social welfare measure for the valuation of transport time changes in the system. We limit ourselves to the area of freight transport, while taking into account not only transport services but the wider, transport demand-related background of freight logistics.

1.2. The relevance of transport time to the productivity of logistics systems

If one wants to understand the real value of transport time to the economy, we argue that it is of crucial importance to consider the wider context of logistics, production and trade activities, through which time acts as a resource. Freight transport demand is a derived demand, resulting from the spatial interaction between many complex industrial processes. The main linkage between transport processes and the wider economy is the activity of supply chain management, providing the organisation of movements of goods between firms. When discussing the importance of freight transport time, or changes therein, to the economy, we should therefore first ask ourselves: how important is time for logistics activities and, in conjunction with these, for various other activities of industry, such as trade, marketing and production?

Clearly, for the manufacturing and service industries, time is an important resource, firstly, as a cost driver in production and logistics processes: drivers' wages can account for over 60 per cent of the primary transport costs. Secondly, and perhaps more importantly, it has become a critical element in the competition between firms for customers[1]. As Davis (1987) puts it[2]: *"Customers use our time until their decision to buy, after that we are using their time. (...) The key is the shortening of the elapsed interval between the customer's identified need and his, her or its fulfilment."* This pressure on performing well for the supply chain as a whole translates to the transport system. The other way round, it is clear that the value of transport time is directly related not so much to the transport sector itself but to the quality of service and products that can be delivered to customers.

As regards transport policy analysis, one must also consider that it is the effect of changes in transport time which should be studied -- in other words, do the industry and its customers reap the benefits of improved transport speed? Naturally, there are many influences which will determine whether time gains really imply an increase in productivity. In contrast to the quotation in the previous paragraph, there are also signs that tend towards a weak linkage between transport times and productivity.

Many firms do not know in which market segments they make a profit and where they lose; only firms in very specific markets feel that congestion threatens their market position. These tend to be firms which have only few options to move their shipments outside congested periods, due to the existence of time windows for deliveries (city distribution) or a long shipping distance towards customers with competition based on time-to-market (e.g. flower business). Some of these companies adapt the speeds in their route planners with -10 per cent yearly to account for general congestion trends; this seems to imply an enormous rate of productivity loss. As this is only a small portion of the industry, however, possible time gains through a successful policy will only apply to a limited part of the business population. Also, depending on the transport distances, loading and unloading can take up to 50 per cent of transport time, which makes a slight improvement in travel time relatively unimportant.

Again, we conclude that in order to evaluate the importance of time gains it is necessary to look deeper into the importance of these logistic activities to the wider economy.

1.3. Objectives and structure of the paper

The objectives of this paper are to provide:

1) a summary of the state of the art in transport research related to freight VOT;
2) a conceptual overview of sources of economic benefits through time gains;
3) proposals for research to improve our knowledge of freight VOT.

Chapter 2 summarizes the state of the art in research directed at valuation of logistic improvements, provides an overview of actual estimates reported in the literature and discusses some methodological limitations of VOT research. Chapter 3 discusses the relevance of recognising the logistics perspective for Cost-Benefit Analysis and proposes ways to extend present research practice. We close our paper in Chapter 4 with some brief concluding statements.

2. THE VALUE OF LOGISTICS IMPROVEMENTS: THE CURRENT PARADIGM

2.1. Methodological background

2.1.1. The basic approach

We take the method of Cost-Benefit Analysis (CBA) as the context in order to clarify how the value of logistics improvements is measured at present. The motive for considering logistics improvements in CBA is straightforward. Ultimately, CBA aims at showing whether an investment will result in an improvement in welfare. The traditional criterion applied to assess this is the so-called Hicks-Kaldor criterion, in terms of which an investment is justified if those persons who gain from it can compensate those who lose from it. The approach used to determine whether or not the criterion is met is to assess what is referred to as the willingness to pay or, more formally, the compensating variation (CV).

The resources tied down in transport and storage are not ends in themselves; they are inputs into final consumption. Therefore, if an investment results in logistics improvements, resources are freed up which can be used elsewhere in the economy and to produce more goods and services desired by consumers. The ambition of CBA is to measure the willingness to pay for being able to obtain these additional goods and services, made possible by way of the improvement in infrastructure. The question is: how is willingness to pay to be measured?

The textbook approach is based on the assumption that the demand for the good subject to transport and storage can be explained by way of a demand function. This function indicates how demand is determined by a set of factors (variables), including the time required to transport the good from A to B and the quality of this transport. It is furthermore assumed that the quality dimension may be represented by measurable variables reflecting the uncertainty in transport time, the risk of accidents, and loss and damage en route and whilst loading and unloading. Given that (i) such a function exists, (ii) the actual levels of the explanatory variables before the investment is incurred may be determined and (iii) their levels after the investment has taken place may be determined as well, then this function can be used directly to determine the willingness to pay (the CV) for the investment. Such a demand function would, in other words, enable the analyst to directly determine the economic value of the logistics improvement. A number of assumptions have to be fulfilled for the approach to be valid, including perfect competition.

Textbook demand functions only exist in textbooks. In real life, much simpler demand functions have to be made use of. The basic idea of the approach used in practice is that the total logistics cost of a particular shipment can be decomposed into, in principle, three basic components:

–   The cost of transport (e.g. the cost of using a goods vehicle);
–   The cost of "freight" (reflecting that goods in transit cannot be consumed);
–   Other costs, such as damage and uncertainty on account of transport times not being fixed (the quality factors).

It is further assumed that the last two of these components in turn can be expressed in a format which is referred to as a generalised cost. A generalised cost is based on the idea that the cost can be formulated as the sum of the product between a variable (assumed to influence demand) and a unit value, a price. Generally, the cost of transport is assumed to be a part of the generalised cost of the shipment, although it is not necessary to make such an assumption. When the transport cost is part of the generalised cost it is usually assumed that a transport improvement, resulting in reduced logistics costs and increased demand for shipments, will not affect the size of the shipment.

The approach used in practice is based on the following further assumptions:

–   Unit values are constant with respect to the level of each variable entering into the generalised cost function (and explaining demand);
–   Unit values reflect willingness to pay (measured in the form of a CV).

For the purposes of a CBA of a transport intervention, it is thus conventionally assumed that the valuation of logistics improvements may be decomposed into a transport element and a goods element, and that the costs associated with the goods, including the time required for their shipment and other quality factors, can be examined separately.

The goods component of the logistics cost comprises the following cost elements (as is also borne out by the available studies into the goods component, which will be reviewed below):

– The cost of goods in transit whilst being transported from the location of production to the location of its use (consumption);
– The cost of the uncertainty (or unreliability) of the duration of the time for transport, i.e. that it normally cannot be assumed that the transport time is fixed;
– The degradation, loss and damage of the goods in transit from production to consumption.

The unit values to be applied to calculate these costs may be referred to as:

– The (unit) *value of freight time* (typically applied on the *expected* time or the *expected* time saving;
– The (unit) *value of (improved) reliability* (rarely applied in practice and estimated in different ways; see below);
– The (unit) *value of (reduced) damage* (rarely, if ever, applied in practice; see below).

In the sequel to this section of the paper, we focus on the goods component -- being part of the total logistics cost -- and its associated unit values.

2.1.2. Methods for measuring unit prices

As will be discussed further below, there are two basic approaches for determining unit values. One, which is based on market prices, is the same as is used when determining, for example, vehicle operating cost (VOC) savings in a CBA.

The other approach is based on inferring values from choices made between available alternatives for undertaking a shipment, e.g. a choice between different routes or modes. Such alternatives may each be associated with different transport times and quality levels, i.e. levels of the explanatory variables. The econometric models used are typically logit models, which express the probability of the choice of an alternative in terms of the generalised cost of that alternative as well as of the other available alternative(s). To estimate the parameters of the econometric model, there is a need for a set of observations on (i) the actual choice made, and (ii) the level of the explanatory variables which enter into the generalised cost functions.

There are two methods for obtaining data sets of this nature, allowing for the deduction of unit values by way of econometric models: data on observed choices [revealed preference (RP) data], and data on hypothetical choices obtained through interviews [stated preference (SP) data]. In a few studies, a mixture of revealed and stated preference data have been used to estimate unit values.

Stated preference data are typically obtained through interviews involving games. The respondent is presented with sets of data on explanatory variables and is asked to identify the alternative which he/she prefers. The levels of the explanatory variables are related to a real situation, but the choices made in the interview are based on hypothetical levels. These hypothetical data may be obtained through identification of alternatives that the respondent could face in real life. Often, however, they cannot be related to a real-life situation and they are then said to involve abstract choices.

Most of the research on the cost of the goods component of transport has focused on determining unit values based on SP and RP data (see Tables 1 and 2), but unit values used in practice have partly been derived by using the market price approach and partly the approach based on SP data.

2.1.3.   The unit price approach in practice

Two previous surveys [Aaltonen (1993) and Waters *et al.* (1995)] have concluded that the impact on goods is not being taken into account in the CBA methodology applied by the road (and where relevant, rail) authorities in different countries. As a whole, that still applies although Sweden is an exception since at least the early 1980s. Unit prices related to goods are thus made use of in the CBA methodology applied by the National Roads Authority (VV) and the National Rail Authority (BV). In addition, such unit prices feature in the SAMGODS model system, which is a model for forecasting goods transport by different modes in Sweden, including import, export and transit traffic.

Research into the cost of goods in transport has been carried out in a number of countries and recommendations have been made but, so far, no action has been taken by the authorities concerned. This applies in particular to Holland and the UK, countries in which large studies were carried out in the mid-1990s on behalf of the Ministry of Transport and the Department of Transport, respectively[3]. However, the Department for Transport (as it is now called) is apparently currently reviewing the matter[4].

In the context of the European Union, it is noted that values of freight time have been proposed by EUNET (1998), a project funded by the European Commission under the Transport RTD Programme of the Fourth Framework Programme. The purpose of EUNET is to develop a methodology to evaluate socioeconomic effects of transport infrastructure investments. Deliverable D9 provides "freight user values of time" (i.e. unit values of freight time). The report notes that Sweden is the only country to make use of such values, and hence recommends that the values used by Sweden also be used by other EU countries, suitably modified to the cost level in these other countries. In a report from the Western European Road Directorates (2000), recommendations are made for use of the EUNET freight values of time.

Austroads, the association of road authorities in the states of Australia as well as New Zealand, publishes estimates of freight travel time values for a range of freight vehicle stereotypes [Austroads (2000)]. Austroads has commissioned further research and additional recommendations are expected to be forthcoming in the near future[5].

International financing institutions, such as the World Bank (IBRD), the European Bank for Reconstruction and Development (EBRD) and the European Investment Bank (EIB), which apply CBA when appraising road schemes, do not normally account for logistics improvements, except costs related to vehicles and their drivers. The Highway Development and Management Model (HDM-4), a comprehensive tool for analysing road schemes which is frequently used by these institutions and their clients, provides for the calculation of "cargo holding costs" by multiplying with an hourly value of freight time. The analyst has to provide the freight value of time per road transport vehicle. This function is rarely, if ever, employed by HDM users.

2.2. Empirical estimates based on the market price approach

2.2.1.   The value of freight time

As will be elaborated upon below, SP studies yield values of time and reliability which are higher or much higher than values obtained through using the approach based on market prices. In Sweden, BV and VV at present make use of values of freight time which are based on the latter approach. This is often also referred to as the capital value approach, which expresses the unit price in a monetary unit per time unit saved (e.g. hours).

There are three assumptions with this approach:

- The rate of interest;
- The value of the goods being carried per tonne, including the number of tonnes being carried per vehicle unit;
- The number of hours per year.

It is to be emphasized that a value of freight time based on the capital value approach only reflects the savings made from goods being able to reach their destination quicker, thereby reducing the working capital invested in the goods. Another way of looking at this saving, and from the point of view of the entire time period of a CBA, is to recognise that goods being transported are, in effect, goods which are being stored. The discounted time savings in the CBA hence reflect the goods which will be released from this stock and made available to consumption.

It is to be emphasized that this interpretation of what is actually reflected in the value of a freight time saving, estimated in terms of the capital value approach, does not in any way depend on whether demand and/or supply is uneven in time, or production or transport only taking place during a given number of hours of the year (e.g. during 3 600 hours, as conventionally assumed). The value of freight, in terms of the capital value method, always reflects the condition that the time required for transport implies that goods are being stored on a vehicle. A time saving during a transport operation (repeated continuously during the period of the analysis) therefore gives rise to a saving in the stock required to bridge the location of where the good is produced and the location where it will be used.

It is sometimes argued that unit values determined on the basis of the capital value approach do not reflect the full willingness to pay for goods to arrive more quickly at their destination. The argument is related to the condition that demand for the goods in question is stochastic; this should not be confused with the condition that the transport time may be uncertain, a matter to be considered in the next chapter.

One can think of several explanations of why stochastic demand may give rise to a high willingness to pay for quick delivery. One example is when an accident occurs, requiring urgent transport of, for example, spare parts to enable a production process to continue, which otherwise would have to stop in the absence of stocks. Another example is the nature of modern production and logistics methods, which often makes it cheaper not to meet a demand immediately through an available stock, but by receiving an order and then producing the good that is demanded. Of course, once an order has been placed, the customer and the seller often want immediate delivery, and there may therefore be an additional willingness to pay for this.

It is recognised that, in both these cases, there is a possibility that the customers' willingness to pay would be higher than the value reflected through a straightforward application of the capital value approach. As concerns the first example, it may be argued that it could not be a very common feature in relation to total transport flows, but also that the additional value is limited. The reason for saying this is that the person responsible for the production process made the decision not to stock the required spare parts after all, instead relying on delivery in case of a breakdown. Apparently, this latter alternative was viewed as being less expensive.

As concerns the second example, it may be argued that if there is a high demand for immediate delivery, then the market will meet this demand. Indeed, many products may today be bought either through an order against later delivery or directly off the shelf, reflecting the fact that some consumers are prepared to pay a premium for immediate delivery. But this, of course, also means that those who are prepared to wait are those who, for various reasons, can wait. It is difficult to find an explanation of why those who are prepared to wait would be willing to pay more for reduced time in transit, on account of an intervention in the transport system, than what is reflected through the capital value approach.

2.2.2. The value of improved reliability

The capital value approach can be taken further by also accounting for reliability and damage. A theoretical analysis of the issue of reliability has been presented by Minken (1997) and a simplified version is to be found in Bruzelius (1986). However, as far as is known, no extensive empirical analysis has so far been made based on this approach. It would require studies of the variation in the transport time, which is seen as the source of unreliability. The approaches used by Minken and Bruzelius both presume that the variability in transport time can be described in terms of a probability density function and that logistics planners build up stocks to ensure that stockout will not occur, or will occur rarely on account of the variability in transport time. The value of improved reliability is determined from the reduction in the buffer stock made possible by a reduction in the variability in transport time. The method may be used to obtain values of reliability through simulation. The attractiveness of the approach is that it allows for expressing improvements in reliability in terms of variables which may be measured in the context of a CBA of a transport intervention, e.g. by way of changes in the standard deviation in the transport time. A further property of the approach is that it can be used to provide estimates which may be seen as an upper limit on the value of improved reliability, in view of the fact that logistics planners have a choice between using a buffer stock and not using such a stock at all. When the latter alternative is chosen, it is cheaper to allow for a stockout.

In the Bruzelius study, a simple model is used in order to obtain an estimate of the value of reliability. It is based on the assumption that arrival times are normally distributed and that the relationship between speed and the variation in transport time is linear. It is shown that, when applying the rule that stocks should be adequate 99 per cent of the time, the value of improved reliability may be assumed to be somewhat higher than the value of freight time, i.e. the size of the reduced stock on account of improved reliability is somewhat higher than the stock reduction on account of reduced transport time. Bruzelius argued that his model would provide a reasonable estimate of the value of reliability in CBA of many normal road investments for the following reasons. The model would be likely to yield an overestimate because:

- the specific assumptions made with respect to the relationship between speed and the variation in transport time;
- stocks are not only maintained on account of the variability in transport time but also because of variability in demand;
- only part of the transported goods is subjected to stringent arrival times requiring the build-up of stocks.

The model might, on the other hand, result in an underestimate on account of:

- the actual cost of maintaining stocks (i.e. the cost of the warehouse and warehousing);
- the variation in arrival times might be better described by a skewed distribution (e.g. the log-normal, see below), which could result in the need to hold a larger stock at a given target level for a stockout, say stocks available at 99 per cent of the time.

The conclusion of Bruzelius (1986) was that it would be reasonable in CBA of normal road investment to account for reliability resulting in reduced variability in transport time by doubling the value of freight time (based on the capital value approach). Bruzelius also argued that the detailed measurement of the value of reliability should not be seen as important, in view of the fact that the cost of freight time normally plays an insignificant role in comparison with other components in a CBA.

2.2.3. The value of reduced damage

No attempts appear to have been made to determine the value of reduced damage using the capital value approach. This approach would require information about:

–   the value of goods (data are available);
–   the risk of damage (and loss) per km or per hour for different modes;
–   the nature of the damage, i.e. the proportion of the goods that would not be accepted and would have to be disposed of on account of damage.

The value of damage determined in this way cannot be expected to be substantial, although this is a matter which needs to be investigated further. The INREGIA (1999) study suggests that 11 per mille of goods transported by road and 22 per mille of goods transported by rail are subjected to damage. These values refer to the entire transport operation from start to end, and presumably reflect the loading and unloading operations as well. They can therefore not (given the information available through the report) be converted into values per hour or km. In addition, it is unclear what is meant by damage in the INREGIA study, in that it cannot be determined what proportion of a shipment, which has been described as having being damaged, has actually been fully lost.

2.3. Empirical estimates based on stated- and revealed-preference analyses

The second approach to determining values for freight time, reliability and damage is by way of revealed- and stated-preference analysis. As evidenced by Tables 1 and 2, there is a large number of analyses of this nature.

2.3.1. Values of freight time

Table 1 presents the values of freight time obtained through SP and RP studies. Note that the included values reflect different currencies, years and units. The last column also indicates if the estimates have been obtained from data on decisions made by road transport companies (operators) or models which have not been formulated in terms of linear generalised costs (non-linear). To provide a reference point: the current values of freight in Sweden, derived from the capital value approach, are on average SEK 35 for a road transport vehicle and SEK 23 for a rail wagon. Whilst being overestimated (in terms of the capital value approach)[6], they appear to be low in comparison with the values obtained through the SP and RP approaches.

Table 1. Results of SP and RP studies: Values of freight time

Study	Year of data	Mode/ Country	Value	Unit value per:	Comment
Transek (1990)	1989-90	Rail/S	SEK 6	hour & wagon	Non-linear
Transek (1990)	1989-90	Road/S	SEK 20	hour & shipment	Non-linear
Transek (1992)	1991	Road/S	SEK 30	hour & shipment	Non-linear
Kurri et al. (2000)	1997	Road/SF	$ 1.53	hour & ton	
Kurri et al. (2000)	1998	Rail/SF	$ 0.1	hour & ton	
Fridstrøm et al. (1995)	1992	Road/N	NOK 0-70	hour & shipment	Non-linear
Hodkins et al. (1978)	1970s(?)	Road/Sea/AUS	AUS$ 10	day & ton	RP
Kawamura (2000)	1998-99	Road/US	$ 23.4-26.8	hour & shipment	Operators
Wigan et al. (2000)	1998	Road/AUS	AUS$0.66-0.40	hour & pallet	
Wynter (1995)	1990-94?	Road/F	FF 7	min.& shipment	Operators
De Jong et al. (2001)	2000	Road/F	FF 29-60	hour & shipment	SP+RP
"	2000	Rail/F	FF 17-73	"	"
"	2000	Combined/F	FF 34-53	"	"
Fosgerau (1996)	1988-89	Road/DK	DKK 2.7-6.0	min. & shipment	Operators
Winston (1981)	1975-77	Road/US	$125-1187	day & shipment	RP
Winston (1981	1975-77	Rail/US	$490	day & shipment	RP
De Jong et al. (1992)	1991-92	Road/NL/ (99 prices)	$ 32-42	hour & shipment	Non-linear
"	"	Rail/NL/ "	$ 32	hour & wagon	"
"	"	IWT/NL/ "	$ 222	hour & shipment	"
Fowkes et al. (2001)	2000-01	Road/UK	£ 37.2-169.3	hour & shipment	Partly operators
De Jong et al. (2000)	1994-95	Road/UK/ (99 prices)	$ 21-48	hour & shipment	Partly operators
Fowkes et al. (1991)	1988-89	Road/UK / 99 prices	$ 0.09-1.29	hour & ton	
Viera (1992)	1990?	Rail/US/99 prices	$ 0.59	hour & ton	SP+RP
Roberts (1981)	1980?	IWT/US/99 prices)	>$ 0.05	hour & ton	RP
Blauwens et al. (1988)	1985?	IWT/B/99 prices)	$ 0.1	hour & ton	RP
Fehmarn Belt (1999)	1997?	Road/DK+D/ 99 prices	$ 21	hour & shipment	Operators?
De Jong et al. (1995)	1995	Road/D/99 prices	$ 33	hour & shipment	Non-linear
"	1995	Road/NL/ 99 prices	$ 40-43	hour & shipment	Non-linear
"	1995	Road/F/ 99 prices	$ 34	hour & shipment	Non-linear
Bergkvist et al. (2000)	1991	Road/S	SEK 14	hour & shipment	
Bergkvist (2001)	1991	Road/S	SEK 34-509	hour & shipment	
INREGIA (2001)	1999	Road/S	SEK 0-227	hour & shipment	
"	1999	Rail/S	SEK 0	hour & shipment	
"	1999	Air/S	SEK 117	hour & shipment	
Small et al. (1999)	1995?	Road/US	$ 144-193	hour & shipment	Operators

IWT = Inland waterways transport.

2.3.2. Values of improved reliability

The most common approach to estimating the value of reliability in SP studies is by way of the variable percentage (or per mille) delay. During an SP interview, the first stage involves identifying the number of delayed consignments for a typical transport operation of the respondent's firm. In the second stage, the respondent is then asked to choose between alternatives which involve other values on the delay variable (as well as other influencing variables). In some studies, a distinction is made between goods which are considered delayed if they arrive late during the agreed arrival date, and goods which are considered delayed only if they arrive the day after the agreed arrival date. The theoretical basis for using this specification is unclear.

The recent study by Fowkes *et al.* (2001) suggests an alternative way of measuring reliability, and by way of what is called the spread. The spread is the time between the earliest arrival time of a given shipment and the time when 98 per cent of all shipments have arrived. The spread variable is thus based on the notion of a probability distribution of arrival times, but tries to describe it in a very simple way. However, its usefulness is limited. It will normally not be possible to determine how the spread time is changed through a proposed investment to be appraised with CBA.

In some US studies, attempts have been made to take into account variability through the standard error of the estimated travel time or the coefficient of variation, i.e. the ratio between the standard deviation and the mean travel time. The study carried out by Small *et al.* (1999) assumes that the travel time follows the log-normal distribution, and then attempts to estimate values of reliability reflecting this distribution, using three different approaches: (i) the standard deviation; (ii) coefficient of variation and (iii) a function reflecting that arriving early is associated with a cost per time unit, whilst arriving late gives rise to both a fixed penalty and a penalty per time unit delayed. The number of observations used in this SP study was, however, limited and the respondents apparently also expressed difficulties in understanding the questions posed, including the variables reflecting early and late arrivals, which were formulated so as to reflect the assumption of a log-normal distribution. The usefulness of the results of this study is also limited by the condition that the interviews were carried out with transport operators.

The appropriateness of all the above approaches to measuring reliability may also be questioned, as they are based on the assumption that delay or reliability is not related to the length of the trip or its duration in time. It would seem more reasonable to assume that reliability is a function of the time duration. If, in addition, this relationship could be assumed to be linear, a value on reliability could readily be incorporated into the standard CBA framework. Indeed, if reliability is a function of transport time, one possible explanation for the high values of freight time obtained through SP and RP studies could be that they also reflect the value of reliability.

Some of the results obtained from SP and RP studies are presented in Table 2.

Table 2. **Results of SP studies: Values of reliability**

Study	Year of data	Mode/ Country	Value	Unit value per:	Comment
Transek (1990)	1989-90	Rail/S	SEK 60 same day	1 % unit & shipm	Non-linear
Transek (1990)	1989-90	Rail/S	SEK 40 next day	1 % unit & shipm	Non-linear
Transek (1990)	1989-90	Road/S	SEK 150 same day	1% unit & shipm	Non-linear
Transek (1990)	1989-90	Road/S	SEK 30 next day	1% unit & shipm	Non-linear
Transek (1992)	1991	Road/S	SEK 280 same day	1% unit & shipm	Non-linear
Transek (1992)	1991	Road /S	SEK 110 next day	1% unit & shipm	Non-linear
Kurri et al. (2000)	1997	Road/SF	$ 47.47	hour & ton	Expected delay
Kurri et al. (2000)	1998	Rail/SF	$ 0.50	hour & ton	Expected delay
Wigan et al. (2000)	1998?	Road/AUS	AUS$1.25-2.56	1% unit & pallet	
De Jong et al. (2001)	2000	Road/F	Not reported	1% unit & shipm	SP+RP
"	2000	Rail/F	"	"	"
"	2000	Combined/F	"	"	"
Winston (1981)	1975-77	Road/US	$ 404	day, standard dev.	RP
Winston (1981	1975-77	Rail/US	$299-4110	day, standard dev.	RP
De Jong et al. (1992)	1991-92	Road/NL/	Not reported	1% unit & shipm	Non-linear
"	"	Rail/NL	Not reported	1% unit & shipm	"
"	"	IWT/NL	Not reported	1% unit & shipm	"
Fowkes et al. (2001)	2000-01	Road/UK	£ 61.5-167.6	hour & spread	Partly operators
De Jong et al. (2000)	1994-95	Road/UK	Not reported	1% unit & shipm	Partly operators
De Jong et al. (1995)	1995	Road/D	Not reported	1% unit & shipm	Non-linear
"	1995	Road/NL	Not reported	1% unit & shipm	Non-linear
"	1995	Road/F	Not reported	1% unit & shipm	Non-linear
Bergkvist et al. (2000)	1991	Road/S	SEK 165 same day	1% unit & shipm	
"	1991	Road/S	SEK 84 next day	1% unit & shipm	
Bergkvist (2001)	1991	Road/S	Not reported	1% unit & shipm	
INREGIA (2001)	1999	Road/S	SEK 63	1 per thousand & shipment	From linear model
"	1999	Rail/S	SEK 1142	1 per thousand & shipment	From linear model
"	1999	Air/SWE	SEK 264	1 per thousand & shipment	From linear model
Small et al. (1999)	1995?	Road/US	$ 371.33	hour & shipment	Expected delay, operators

2.3.3. *Values of reduced damage*

A limited number of the studies include variables related to damage and loss, e.g. Transek (1992), de Jong et al. (1992) (the Dutch VOT study), de Jong et al. (1995) (studies of road transport in Holland, France and Germany), Fridstrøm and Madslien (1995), Wigan (2000) and Bergkvist and Westin (2000). With the exception of the study by Wigan (2000), in which damage is defined in terms of pallets not being accepted by the receiver, it is not fully clear what is meant by damage, i.e. whether the whole shipment is lost or only part of it.

Estimates vary significantly. The Transek (1992) study resulted in an estimate of SEK 270 for one per mille unit reduction in the frequency of damage. Bergkvist and Westin (2000), using the same data but a linear specification and another estimator, obtained a value of damage of SEK 20 for a per mille unit reduction in frequency of damage to a shipment.

As in the case of reliability, the SP and RP values for damage are of limited usefulness from the point of view of CBA, as they are not expressed in relation to the distance or the time duration of shipments.

2.4. Methodological issues in existing practice

Traditional empirical studies concerning the goods dimension of logistics costs focus on three aspects: the time duration of transport, reliability and damage and loss. Two approaches may be used to estimate the values associated with these three variables, namely, market prices and using values derived from models of choice between transport alternatives, in which the three variables are explanatory factors.

The two approaches give very different results. The approach based on market prices (also referred to as the capital value approach) results in low prices. Even taking reliability into account is not likely to change that assessment. It appears that the value of reliability using the market price approach would yield values per time unit saved, which are of the same order as the "pure" value of freight time based on the market price approach. No estimates have been prepared for damage and loss using this approach, on account of lack of data on the probability of damage and loss per km or time unit. This is a void which should not be all that difficult to fill.

The SP approach to the estimation of values of relevance to freight, in particular, has attracted much interest during the last decade, and there is an abundance of results. They indicate substantial variability and appear to be very sensitive to the specification of the model used and the method of estimation. The value of these estimates must be viewed as not without problems for the following reasons:

- SP and RP models make use of variables which are not always practical from the point of view of performing cost-benefit analyses of transport schemes. For example, reliability is in these models often measured in terms of the portion of shipments which arrive late. When appraising transport interventions by way of CBA, it would normally not be feasible to measure their impact on the portion of shipments arriving late.

- The SP and RP approaches are based on methods which raise a number of issues concerning what is being measured, and therefore whether estimated values are valid from an economic point of view. There is a need to address such issues through research before the estimates of these models are accepted for use in officially sanctioned CBA methodology.

The following issues may thus be identified for the SP and RP approaches:

- **Heterogeneity.** SP and RP discrete choice models involve choice between alternatives for how to transport *shipments*. These shipments are, of course, very diverse in nature (value, size, weight, distance), whilst the models normally used to explain choice presume a value which is independent of these characteristics. Attempts are often made to lessen the effect of this restriction by estimating separate values for different types of goods, etc. Even under these circumstances, a fixed value of time must be viewed as a very strong assumption. There are examples of studies in which distributions have been introduced for the values of time (e.g. Wynter, 1995 and Kawasaki, 2000), but these are exceptions (and the studies mentioned also suffer from other shortcomings). In other studies, homogeneous commodity groups called "logistics families" are used (Tavasszy, 1998), where variations in the willingness to pay within categories are assumed to be minimized.

- **Non-linearity.** The CBA methodology is based, as mentioned, on the concept of linear generalised costs. However, a number of studies have estimated values from non-linear functions, including the two Swedish studies from the early 1990s (Transek, 1990 and 1992). The validity of estimates obtained from non-linear functions is unclear. The type of function used matters. The data from the 1992 Transek study have subsequently been used by Bergkvist and Westin (2000), who estimated values of freight time by applying a linear function as well as using a different estimator. The value of freight time obtained by Bergkvist and Westin (SEK 14 per shipment and hour) was less than half that obtained through the original approach (SEK 30).

- **Restrictive behavioural assumptions in models (IIA).** The most popular model used to estimate values from SP data is the logit model. The logit model is based on the assumption that the error term of the generalised cost function is independently distributed (with an extreme value distribution). However, SP data are obtained through repeated interviews, which means that the independence assumption is not fulfilled. This results in biased estimates (overestimates) of the t-values and in the significance of parameter estimates being overstated, see Cirillo *et al.* (1996).

- **Object of measurement (i).** This point has strong relations with the issue of completeness of the analysis, raised earlier. Positions held by the person interviewed to obtain SP data vary from study to study. In general, a person in a managerial position, having the authority to make decisions, is interviewed. This person normally represents the shipper (consignor), less often the operator and even less often the receiver (the consignee). The importance of who is actually interviewed is analysed from a theoretical point of view by Winston (1981). An aspect to consider is, of course, the terms of the contract for the purchase of the goods, in particular how these are being paid for. If paid CIF, the shipper is probably the key person to interview, but if the goods are sold FOB, the receiver is likely to be more important. This issue will, of course, not matter when the shipper and receiver come from the same firm (i.e. transport is internal). It would not seem appropriate to use data on choice made by transport operators (i.e. firms which undertake transport for hire and reward) as they will most likely be more concerned with the cost of their own operations than the cost of the goods they carry (although contractual arrangements may impact on this). It is noted that, with the exception of the study by Winston, the issue as to whether the shipper or the receiver should be interviewed is not addressed. It is also noted that estimates of values of time obtained from managers of hauliers tend to be higher than those obtained from respondents representing shippers. See, for example, the study by Fowkes *et al.* (2001) and the study reported on by de Jong (2000), which contains data from the UK value-of-time study in the mid-1990s. The values of freight time for operators in these studies are much higher than the values obtained from shippers[7]. Note that behavioural studies can also give different results depending on the type of shipper who is being observed: own-account shippers will take into consideration the potential multiplier effects of time savings on their logistics, whereas shippers who use third-party services will do this only to a limited extent.

- **Object of measurement (ii).** This last point brings up a related issue, *viz.* to what extent the choices made by SP respondents are clouded by other concerns which do not enter into the alternatives -- in the form of variables -- that they can choose between in the SP games, or by variables being incorrectly specified. Similarly, it may be queried to what extent the respondents in the interviews can consider the longer-term consequences of the alternatives.

It should be emphasized that, from the point of view of a CBA, it is not short-term effects that are of interest but the longer-term impacts, reflecting a situation where the economy has found a new equilibrium after a change.

–   **Experiment design.** More fundamental as concerns the quality of the data is the question of to what extent a respondent is, in fact, able to rank alternatives in terms of a utility function (fulfilling the assumptions typically made, such as completeness, transitivity and reflexivity). During interviews, the respondents normally have limited time before they have to make a choice, and he/she never incurs any rewards or penalties for making a decision. The INREGIA (2001) study suggests that respondents often make a decision on the basis of the level of one variable only instead of weighing together all the variables of each alternative before making the choice. Is it possible that the data obtained through SP interviews are better explained in terms of a lexicographic ordering than the ordering assumed by the economist? The question has not been analysed. It should also be emphasized that the cost of transport is small in relation to the value of the goods (some 2 to 3 per cent in most instances), which may have an additional effect on choice, not only in SP interviews but in real life. To this should be added the fact that the abstract alternatives in SP interviews are often structured in such a way that the respondents can choose between alternatives which imply better quality at a higher cost. Everybody wants better quality, and the decisionmaker who serves as the SP respondent is probably often rewarded for improving quality. If the cost to be paid for improved quality in the game is no real cost, then will the data not tend to overstate the willingness to pay for better quality? The available evidence suggests that the way in which the alternatives are structured have an impact on the answers obtained. In the 1994-95 value of freight study in the UK (reported on in de Jong, 2000), transport operators (for-hire and own-account operators) were subjected to two sets of games, one involving abstract alternatives and one involving a toll road and a toll-free alternative. The latter experiment yielded values in the range of GBP 21-34, whilst the former resulted in values in the range GBP 36-48. It is understood that the same persons were interviewed in the two games.

Many of the above issues relate to SP studies alone. But also the few available RP studies are characterised by data problems. RP data have thus often been obtained *ex post* through reconstruction by the researcher and may therefore not represent the actual prices, times and quality factors faced by the decisionmaker when he made his decision.

In this chapter, we reviewed the present state of the art and provided some comments as to opportunities for an incremental improvement of research practices, to further increase the validity and accuracy of VOT estimates. Notwithstanding the ever-existing methodological questions, the VOT concept is of invaluable importance to CBA practice. The VOT concept, as discussed until now, relates to the costs and benefits that accrue to the first users and the producers of transport services. As such, it provides a building block for CBA of transport policy in the sense of the direct effects of changes in transport service quality. In the next chapter, we will look at how the range of costs and benefits that is considered with the VOT can be extended, to consider also indirect effects of transport policies.

Round Table 127: Time and Transport – ISBN 92-821-2330-8 – © ECMT, 2005

3. WIDENING THE SCOPE OF VOT MEASUREMENTS

3.1. A full account of reorganisation responses?

Does present practice provide a sufficient account of the willingness to pay for changes in transport time? As VOT is usually used in CBA to denote the direct effects of transport time changes (i.e. effects on the transport service provider and the first user), present practice shows two restrictions:

1. Measurements usually do not go beyond changes in the choice of mode. But spatial reorganisation effects do exist and can provide substantial multipliers over the primarily assumed impacts, based on observable market prices. In particular, the impacts of changes in transport time reliability are usually found further downstream in the chain, where inventories need to be minimized or market areas are at stake (Mohring & Williamson, 1969).

2. Additional effects on the wider economy, or further transport-using sectors, are usually neglected. Although the transport sector can hardly overlook consequences for the complete chain of sectors which indirectly depend on the quality of their services, it can be expected that additional savings can be made here as well, which now remain unrecorded.

Both the capital value and the RP/SP approaches are partial approaches as they are practised now, i.e. they take into account a limited number of changes in the transport market. In the case of the capital value method it is always clear which consequences are studied; in the case of the behavioural approaches the boundaries are less clear. For example, shippers may or may not consider possible future claims by their clients for late delivery in their choice of mode. They may or may not consider in their decisions the possibility to alter the spatial and functional structure of their supply chains (e.g. the number and location of warehouses). They may or may not include the savings in drivers' wages in their choice of mode. The fact that such boundaries are not always clear introduces a risk for CBA in terms of incompleteness or double-counting of effects.

A key aspect of improving the practice of VOT studies is to take into account the full range of organisation responses within the system, first of all by clarifying what is meant by "the system". A uniform conceptual framework is needed for a comprehensive treatment of logistics benefits of changes in transport time. This should clarify what needs to be measured and for which purpose. For the daily practice of project assessment using CBA, this extended framework would relax two restrictive (though practical) assumptions, relating to backward and forward linkages, respectively:

– In many classical transport models, traffic is assumed to be inelastic. Clearly, reorganisation can have a substantial effect on the consumption of various production factors or inputs to the industry (among which is transport demand);

–   In transport CBA, we normally assume that changes in transport time do not cause additional benefits in the wider economy due to reorganisation effects. Reorganisation also influences the prices of products and hence can create benefits that are different from price changes which propagate via the transport sector to transport-using sectors.

Recent studies and debates on the question of whether we are making a serious underestimate of benefits with our present way of working (see, e.g., SACTRA, 2001 or IASON, 2002) could now be looked at empirically and from a systems perspective.

3.2. The conceptual framework from a logistics perspective

There are many examples of basic conceptual frameworks which describe the transport system and its processes (see for example, OECD, 1992, for freight]. Our framework would distinguish the following markets: infrastructure, transport services, logistics services, trade and the local production/consumption markets. The markets are closely interconnected, as they provide each other with supply (service levels and prices) and demand (for a certain volume and nature of services or goods). Based on this scheme, we can see three types of responses which will occur as a result of changes in transport times: transport, inventory and production reorganisation. Changes in transport times and prices that occur in one market can propagate through others (e.g. quicker transport which implies more slack in the supply chain and lower delivery times) or can be absorbed (quicker transport which does not increase the slack in the supply chain because of a longer waiting time).

Table 3. **Classification of reorganisation responses on changes in transport times**

	RE-ORGANISATION DECISIONS	COST/SERVICE DRIVERS
TRANSPORT	• change in driver time assignment • change in routes • change in type/number of vehicles • change in starting / arrival times	• vehicle operating/handling costs • costs of inventory in transit • transport costs of (un)reliability • damage and loss during transport • improved service to shipper
INVENTORY	• change of static inventory volumes • change of # intermediate inventories • change of inventory locations • change in replenishment strategy	• warehouse/handling costs • interest costs of stocks • higher responsiveness • improved order accuracy
PRODUCTION	• change in production technology/ops • change of production location • change of number of production sites • change in product ranges	• lower production costs • closer to consumer market • improved information • increased customization

Using this scheme, we can associate costs or benefits with changes in transport time at various levels in the system. We can measure the value of time by an orderly bookkeeping of all these costs and benefits, in such a way that we do not double-count or miss anything. In order not to double-count, we must realise that these markets and responses cannot be seen as fully independent, as some reorganisation benefits will in part be handed over between the different agents in the system (the propagation effect) or only occur as an effect of various simultaneous changes in the system.

Within this system we aim to take into account the full range of reorganisation responses, recognising, firstly, that the "goods value" of time is in essence a "chain value" of time and, secondly, that the focus is not only on impacts on product price but also on logistics service quality and product quality.

From the perspective of application of VOT figures in CBA, we see the following basic possibilities:

–   *To use the market price approach to have a well-observable estimate of the part of the time-related impacts* -- here the challenge is to include as wide a range of responses as possible. As sketched before, this approach has until now covered both vehicle operating costs and some elements of the "goods" component of effects of time changes.

–   *To use a behavioural approach with transport models or a combination of transport, logistics and regional economic models.* This can be based either on SP or RP data, but is in essence a behavioural approach (see Tavasszy and Ruijgrok, 2003). Here, the challenge is to avoid double-counting in the estimates by closely aligning models and/or experiments.

–   *To use both a market price and a behavioural approach, aiming at complementary (and thus additive and independent), partial VOT estimates.* For example, effects of time changes on vehicle operating costs can be estimated quite accurately with the market price approach. Surveys with shippers who use third-party transport service providers can give an idea of the additional "goods"-related effects, if transport prices are assumed not to change, while door-to-door transport times are.

3.3. Research challenges

It is clear where the framework would lead us, in terms of an extension of evaluation practices and implications for research:

1.   Empirical research into the magnitude and direction of inventory and production reorganisation responses related to changes in transport time, across various sectors and countries;

2.   Operational models that are able to forecast system-wide effects across a range of sectors, down to the individual consumer, allowing an unambiguous assessment of welfare effects;

3.   Adapted VOT estimates which, on the basis of a widely supported approach, map the economic importance of transport time (and other time-related transport performance characteristics) for use in CBA;

4.   Clear rules for CBA practice, given the instruments and data available at present, in order to make sure that no benefits of significant importance remain unaccounted for and that no double-counting is done.

We treat these directions for research in more detail below.

3.3.1. Empirical research into logistics reorganisation responses

The first knowledge gap we need to fill to improve our understanding of reorganisation benefits concerns the potential magnitude and direction of these responses under varying conditions for freight transport time and reliability. Not only will this provide us with new ideas for describing the behaviour of the system, but also we will learn about the potential for producing more accurate and comprehensive project assessments. The FHWA's Freight Benefit/Cost White Paper (Federal Highway Administration, 2001) provides a thorough account of the microeconomic mechanisms of reorganisation responses, the available literature on this topic and the possible consequences for CBA. One of its key recommendations is to estimate direct and indirect logistic cost savings from travel time savings and improved reliability of transport, using survey methods for a sample of firms. This recommendation not only implies a great effort for data collection about companies' logistics decisions but also, in order to provide a behaviourally valid description of the system, the design and implementation of mathematical models at the disaggregate or aggregate level. These models can be set up based on the modelling approach and techniques presented in Daganzo (1999).

3.3.2. Building a new generation of operational models

This research would be aimed at the development of new tools to improve our understanding of the relationship between changes in logistics structures and interregional economic equilibria. Already before, but especially since the research of Venables and Gasiorek (1996), Spatial Computable General Equilibrium (SCGE) modelling has arrived as a means to predict welfare effects of transport investment and policies (see, e.g., Bröcker, 1999 and IASON, 2002). Despite the research problems which remain to be solved (see Lakshamanan *et al.*, 2002 and Thissen *et al.*, 2002), there is a continued move towards integrating transport models and CGE approaches into more comprehensive tools for assessment. Beside SCGE modelling, other, perhaps less complex but equally powerful approaches are being used for assessment, which include dynamics, thus allowing for the accounting of benefits in a full CBA (see the System Dynamics approach in ASTRA, 1999, or the regional production function approach in the SASI model, IASON, 2002b). In order to sharpen our insights into future logistic structures and their relationship with economic development, we propose to include the total logistics costs into the SCGE framework, thus giving a better interpretation of what is now -- in SCGE terms -- referred to as "transport costs". The operational modelling of logistic structures can be done according to the lines of the SMILE and SLAM models, which provide a picture of how logistic structures are affected by regional and product characteristics (see Tavasszy *et al.*, 1998).

3.3.3. Adapted VOT estimates

It is likely that the values of time which are recorded using a comprehensive system framework will be different to the values which are used at present, starting from a partial framework. This will depend on the existence and nature of market imperfections in the chain of markets downstream from the transport system. Also, we can expect that there will be tradeoffs between the value of time and the value of reliability (see, for example, Muilerman, 2002). Firms may choose logistic structures that are relatively insensitive to either unreliability or a long transport time -- depending on the spectrum of transport services available to them and the demands of their clients. The new VOT estimates will be a product from a systematic data collection process and the development and application of new, integrative transport/economy models. Beside the directions provided earlier in this paper in subsection 2.2.2, other relevant work advocating more empirical research in this area can be found in McCann (1989) and FHWA (2001).

3.3.4. Rules for CBA

The results of the above research activities will need to be interpreted in terms of their value to the CBA process. In particular with a view to the estimation of indirect effects, care will need to be taken that sources of benefits are administered correctly, avoiding double-counting of benefits. For example, the question of whether logistics cost savings can be transferred to SCGE models to calculate household level (i.e. final) benefits, will depend on the scope of the reorganisation effects considered. As SCGE models typically consider effects on trade and production (quality and quantity), there is a risk of double-counting if the scope of VOT measurements is not restricted to transport and inventory reorganisation. Rules for CBA typically become important in an international setting, when different methods are used in conjunction, considering direct and indirect effects, with various operating standards and procedures (see, e.g., IASON, 2002a, for recent research on this topic).

The research directions described above still leave the door open to choose an approach based on market prices or on mathematical models (irrespective of whether the data concern RP or SP). Clearly, the behavioural approach has the advantage that it can extend the scope of freight VOT beyond directly measurable market prices and provide simple proxies for a comprehensive valuation of effects of time changes. However, its disadvantage is that the empirical validity of any model is under constant dispute -- error sources exist in the statistical formulation, the experimental design and the actual estimation process, to name a few. The best road to follow will therefore depend on the policy environment within which the VOT is applied.

4. CONCLUSIONS: STATE OF THE ART AND RESEARCH CHALLENGES

This paper has given a birds eye's view of the topic of freight value-of-time. We have focussed on providing a detailed summary of existing research results. However, with the aim of providing input to the research agenda of the future for freight VOT, we have also commented on some methodological problems and discussed opportunities to extend the scope of VOT measurements.

We provide an overview of research results in the area of VOT, value of reliability and value of damage and loss. We discuss some methodological issues in present research practice, especially concerning the behavioural models based on SP data, which require explicit attention if one wants to uphold the validity and accuracy of VOT estimates. The alternative, VOT based on market prices, although more transparent is at present more limited in scope due to practical difficulties with observing representative, real-world prices. For both approaches it is of critical importance to define the context (the actors and the costs considered) within which VOT estimates take place. Notwithstanding the ever-present methodological questions, the VOT concept is of invaluable importance to CBA practice. The VOT concept, as discussed until now, relates to the costs and benefits that accrue to the first users and the producers of transport services. As such it provides a building block for CBA of transport policy, in the sense of the direct effects of changes in transport service quality. The indirect effects of time gains or losses seem hardly to have been treated in the literature.

A promising direction for future research, therefore, lies in extending our framework of thinking towards the full range of logistics reorganisation effects which one can expect from changes in the transport system. We can distinguish transport, inventory and production-related reorganisation effects. In principle, both the behavioural and the market price approach lend themselves to this extension. New research can be directed into four areas: 1) data collection to fill the knowledge gap about firm-level responses to changes in transport time and its variability; 2) development of integrative, operational models which build on this empirical research; 3) refined VOT estimates; and 4) refined rules for the use of VOT values in CBA. This expansion of the conceptual framework allows both the behavioural models and the market price approach to be extended. The choice of approach on which to base our VOT estimates will mostly depend on the policy environment.

NOTES

1. Muilerman (2001) describes in great detail how time-based competition has grown in importance during the last decade.

2. *Ibid*. pp. 23-24.

3. The results have been used on an ad hoc basis in some CBA.

4. No reference is made to the issue in the recent review report on the value of time; see Mackie *et al.* (2003).

5. Mention should be made that TransFund, which funds, *inter alia*, road infrastructure in New Zealand, has been carrying out surveys on the impact on goods of transport interventions. These surveys have, however, not been used to derive values related to logistics improvements with respect to goods; see Melsom (2003).

6. The assumptions made with respect to, *inter alia*, the discount rate and the number of hours per year in effect imply a signficant overestimate in terms of the capital-value approach.

7. Indeed, in his review of the Accent/HCG study, undertaken in 1994 and reported on in de Jong *et al.* (2000) (see Table 1), Fowkes (2001) states the view that the value of the time saving is to be seen as determined by the reduced driver's wages and vehicle operating costs (p. 2).

BIBLIOGRAPHY

Aaltonen, Pekka, Simo-Pasi Antti Permala and Hanna-Kaisa Salminen (1993), Logistics effects do not guide transport network investments, *Nordic Rail & Transport Research,* No. 2, pp. 16-17.

Abdelwahab, Walid and Michel Sargious (1992), Modelling the Demand for Freight Transport, *Journal of Transport Economics and Policy,* January, pp. 49-70.

ASTRA (1999), *Design and specification of a system dynamics model,* ASTRA Deliverable 2, Website: http://www.iww.uni-karlsruhe.de/ASTRA

Austroads (2000), *Economic Evaluation of Road Investment,* Report 142/00.

Bergkvist, E. and L. Westin (2000), *Regional Valuation of Infrastructure and Transport Attributes in Swedish Road Freight,* Umeå Economic Studies No. 546, Umeå.

Bergkvist, E. (2000), *Estimating values of time and forecasting transport choices in road freight with a non-linear profit specification. The logit versus neural networks,* Umeå Economic Studies, No. 540, Umeå.

Bergkvist, Erik (2001), *Freight Transportation; Valuation of Time and Forecasting of Flows,* Umeå Economic Studies, No. 549, Umeå.

Bergkvist. E. and P. Johansson (1997), *Weighted derivative estimation of quantal response models: Simulations and applications to choice of truck freight carrier,* Umeå Economic Studies, No. 455, Umeå.

Blauwens, G. and E. van de Voorde (1988), The valuation of time savings in commodity transport, *International Journal of Transport Economics,* 15, pp. 77-87.

Bröcker, J. (1999), *Trans-European Effects of "Trans-European Networks": A Spatial CGE Analysis,* Mimeo, Technical University of Dresden, Dresden.

Bruzelius, Nils (1986), *Företagens MA-kostnader och företagens kapitalkostnader för fordon, Två uppsatser om samhällekonomiska effekter av vägåtgärder,* PPp Meddelande, No. 1, Statens Vägverk, Borlänge, 1986.

Cirillo, C., A. Daly and K. Lindveld (1996), Eliminating the bias due to the repeated measurements problem in SP data. Paper presented to PTRC Transport Conference, London, 1996.

Davis, S.M. (1987), *Future Perfect,* Addison-Wesley, Reading.

Daganzo, C.F. (1999), *Logistic Systems Analysis,* third revised and enlarged edition, Springer-Verlag, Berlin.

EUNET (1998), *Socio-Economic and Spatial Impacts of Transport*, Deliverable D9: Measurement and Valuation of the Impacts of Transport Initiatives, Institute for Transport Studies, University of Leeds, December.

Federal Highway Administration (2001), *Freight Benefit/Cost Study White Paper; Benefit-Cost Analysis of Highway Improvements in Relation to Freight Transportation: Microeconomic Framework*, FHWA.

Fehmarn Belt Traffic Consortium (1999), *Fehmarn Belt traffic demand study*, Danish and German Ministries of Transport, FTC, Copenhagen, Final Report, 1999.

Fosgerau, M. (1996), Freight traffic on Storebælt fixed link. Paper presented at the 24th European Transport Forum.

Fowkes, A.S. (2001), *Values of Time for Road Commercial Vehicles*, ITS Working Papers 563, Institute for Transport Studies, University of Leeds, December 2001.

Fowkes, A.S., C.A. Nash and G. Tweddle (1991), Investigating the market for inter-modal freight technologies, *Transportation Research, A,* 25(4), pp. 161-172.

Fowkes, A.S., P.E. Firmin, A.E. Whiteling and G. Tweedle (2001), Freight Road User Valuations of Three Different Aspects of Delay. Paper presented at the European Transport Forum.

Fridstrøm, Lasse and Anne Madslien (1995), *Engrosbedrifters valg av transportøsning*, TØI report 299/1995, Oslo.

Hodkins, K.E. and D.N.M. Starkie (1978), Values of Time in Long-distance Freight Transport, *The Logistics and Transportation Review,* 14(2), pp. 117-26.

IASON (2002a), Project Assessment Baseline, IASON Deliverable 1, Website: http://www.inro.tno.nl/iason

IASON (2002b), Methodology for the assessment of spatial economic impacts of transport projects and policies, IASON Deliverable 2, Website: http://www.inro.tno.nl/iason

INREGIA (2001), *Tidsvärden och transportkvalitet -- INREGIA studie av tidsvärden och transportkvalitet för godstransporter 1999*, Underlagsrapport till SAMPLAN, 2001:1, February.

de Jong, G.C., M.A. Gommers and J.P.G.N. Klooster (1992), Time valuation in freight transport: Methods and results. Paper presented at the XXth Summer Annual Meeting, PTRC, Manchester.

de Jong, G.C., Y. van de Vyvere and H. Inwood (1995), The value of freight transport: A cross-country comparison of outcomes, World Conference on Transport Research, Sydney.

de Jong, Gerard (2000), Value of Freight Travel-Time Savings, in: Hensher, D.A. and K.J. Button (eds.), *Handbook of Transport Modelling,* Elsevier.

de Jong, Gerard, Carine Vellay and Michel Houée (2001), A Joint SP/RP Model of Freight Shipments from the Region Nord-Pas de Calais. Paper presented at the European Transport Conference.

Kawamura, Kazuya (2000), Perceived Value of Time for Truck Operators, *Transportation Research Record*, 1725, pp. 31-36.

Kurri, Jari, Ari Sirkiä and Juha Mikola (2000), Value of Time in Freight Transport, *Transportation Research Record*, 1725, pp. 26-30.

Lakshamanan, T.R., W.P. Anderson (2002), Transportation Infrastructure, Freight Services Sector and Economic Growth. White Paper prepared for the US DOT/FHA, CTS, Boston University.

Mackie, P.J. *et al.* (2003), Values of Travel Time Savings in the UK -- Summary Report. Report to the Department for Transport, Institute for Transport Studies, University of Leeds, January 2003.

McCann, P. (1998), *The Economics of Industrial Location, A Logistics Cost Approach*, Springer Verlag, Heidelberg, New York.

McFadden, D.L., C. Winston and A. Boersch-Supan (1985), Joint estimation of freight transportation decisions under non-random sampling, in: E.F. Daughety (ed.), *Analytical studies in transport economics*, Cambridge University Press, Cambridge.

Melsom, Ian (2003), Review of the Benefit Parameters Used in the Transfund New Zealand's Project Evaluation Procedures. Information Paper, Transfund New Zealand, February 2003.

Minken, Harald (1997), *Næringslivets nytte av raskere og mer pålitlig godstransport: Metodgrunnlag*, TØI report 347/1997, Oslo.

Minken, Harald (1997), *Produksjonsmodeller og kostnadsfunksjoner i godstransport*, TØI report 1057/1997, Oslo.

Mohring, H., H.F. Williamson (1969), Scale and Industrial reorganization economies of transport improvements, *Journal of Transport Economics and Policy*, 1969.

Muilerman, G.-J. (2001), Time-based logistics, an analysis of the relevance, causes and impacts, PhD Dissertation, Delft University Press, Delft.

OECD (1992), *Advanced Logistics and Road Freight Transport*, Paris.

Ogwude, I.C. (1990), Estimating the Modal Choice of Industrial Freight Transportation in Nigeria, *International Journal of Transport Economics*, Vol. 17, pp. 187-205.

Ogwude, I.C. (1993), The Value of Transit Time in Industrial Freight Transportation in Nigeria, *International Journal of Transport Economics*, Vol. 20, pp. 325-337.

Östlund, Bo, Christina Bolinder, Gunnar Jansson and Mats Andersson (2001), S*amhällsekonomisk värdering av järnvägsinvesteringar för godstrafik*, TFK Rapport, 2001:1.

Roberts, P.P. (1981), *The translog shipper cost model*, Center for Transportation Studies, MIT, Cambridge, MA.

SIKA (1999), *Översyn av samhällsekonomiska kalkylprinciper och kalkylvärden på transportområdet*, ASEK, SIKA Report 1999:6, June.

Small, Kenneth, Robert Noland, Xuehao Chu and David Lewis (1999), *Valuation of Travel-Time Savings and Predictability in Congested Conditions for Highway User-Cost Estimation*, Report 431, National Cooperative Highway Research Program, Washington, DC.

Tavasszy, L.A., M.J.M. van der Vlist and C.J. Ruijgrok (1998), A DSS for modelling logistics chains in freight transport systems analysis, in (1): *International Transactions in Operational Research*, Vol. 5, No. 6, pp. 447-459; in (2): A.C. McKinnon, K. Button and P. Nijkamp (eds.) (2002), *Classics in Transportation Analysis: Transport Logistics*, Edward Elgar, ISBN 1 84064 551 2.

Transek (1990), *Godskunders värderingar*, Banverket Rapport 9, 1990:2.

Transek (1992), *Godskunders transportmedelsval*, VV 1992:25, October.

Venables, A.J. and M. Gasiorek (1996), Evaluating Regional Infrastructure: A Computable Equilibrium Approach, Mimeo, London School of Economics, UK.

Viera, L.F.M. (1992), The value of service in freight transportation, PhD dissertation, MIT, Cambridge, MA.

Waters, W.G., C. Wong and K. Megale (1995), The Value of Commercial Vehicle Time Savings for the Evaluation of Highway Investments; a Resource Saving Approach, *Journal of Transportation Research Forum*, Vol. 35, No. 1.

Watson, Peter L., James C. Hartwig and William E. Linton (1974), Factors that Determine Mode Choice in the Transportation of General Freight, *Transportation Research Record*, 1061, 138-144.

WERD (2000), *Valuing the Cost and Benefits of Road Transport; Toward European Value Sets*. Prepared for WERD by the Norwegian Public Road Administration, Report No. 1-2000, October.

Wigan, Marcus, Nigel Rockliffe, Thorolf Thoresen and Dimitris Tsolakis (2000), Valuing Long-Haul and Metropolitan Travel Time and Reliability, *Journal of Transportation and Statistics*, December, 83-89.

Wilson, F.R., B.G. Bisson and K.B. Kobia (1986), Factors that Determine Mode Choice on the Transportation of General Freight, *Transportation Research Record*, 1061:26-31.

Winston, C. (1981), A disaggregate model of the demand for intercity freight, *Econometrica*, 49: 981-1006.

Wynter, L.M. (1995), The Value of Time of Freight Transport in France; Estimation of Continuously Distributed Values from a Stated Preference Survey, *International Journal of Transport Economics*, 22:151-65.

Wynter, L.M. (1995), Stated Preference Survey for Calculating Values of Time of Road Freight Transport in France, *Transportation Research Record*, 1477:1-6.

SUMMARY OF DISCUSSIONS

Andreas KOPP
Chief Economist
OECD/ECMT JTRC

SUMMARY

1. INTRODUCTION

The valuation of time requirements for transport and time savings, as a consequence of transport policy action, are often decisive for the acceptance or rejection of transport policies in general and transport infrastructure investment projects in particular. Time savings in passenger transport usually account for about four-fifths of the non-monetary benefits of transport policy measures. The Round Table not only discussed the valuation of passenger travel time but also the valuation of time in freight transport. With the dramatic increase in international trade, the value of time for international freight transport has become an important determinant of the geography of trade flows and of firms' decisions on international location.

An extensive literature exists on the valuation of passenger time. The background paper on passenger time valuation (Crozet, 2005) and the Round Table interventions showed that the interest in the evaluation of passenger travel time is strongly associated with its consequences for our thinking about travel behaviour and the evaluation for transport policies. The unexpected results of empirical research have recently led to a resurgence of research on the value of time in transport (e.g. Mokhtarian, 2005). Standard economic concepts would suggest that the value of time is dependent on travellers' real incomes. In contrast to expectations of increasing values of time in transport and an increasing demand for speed with rising incomes, travel behaviour and stated preferences suggest otherwise.

Much less researched is the value of time in freight transport. This lack of attention, and the limited role which freight transport time savings play in transport policy evaluation, may be based on the view that the effects of transport policy on the freight transport sector can be estimated using market valuations in terms of demand and cost functions. Such an argument overlooks the fact that transport policies often lead to secondary effects which alter cost and demand functions. These changes are due to changes in the reliability of the freight transport system and in network and size economies in the logistics sector, to new regional specialisation patterns and to the relocation of manufacturing industries (Tavasszy, 2005).

Given the recent intensification of international economic relations, the secondary effects increasingly determine the structure and geography of international trade. As was pointed out in the Round Table discussions, the increased importance of transport costs for international trade relations, in view of reduced trade policy barriers, is matched by a shortening of product cycles in manufacturing industries, leading to an increase in demand for speed in international transport. This increased demand for speed may lead to swift relocations of production, with the consequence of rapidly changing international trade patterns (Deardorff,). Transport policy has to meet the challenge of taking account of these changing patterns, and the Round Table discussed how transport policy should react to these changes. Central to the responses is a strengthening of the international co-ordination of transport policies.

2. TIME AND PASSENGER TRANSPORT

As pointed out in the introduction, the valuation of time in passenger transport has received far greater attention in the research literature, as well as in passenger transport policymaking, than has time evaluation in freight transport. The enormous importance of value-of-time savings for transport policy projects and programmes has not led to universally accepted evaluation conventions. One reason for the ongoing debate may be seen in the weak conceptual or theoretical foundations of the value of time in passenger transport; another reason is the difficulty of empirically assessing the value of time. As the Round Table background paper on the evaluation of passenger time points out (Crozet, 2005), problems and hypotheses about the reactions to cost reductions in passenger transport have led to concerns about the negative consequences of the development of the transport sector on the environment, in particular urban development. These policy concerns sometimes seem to feed back to the discussion of how to value the time passengers spend in transport.

2.1. The microeconomics of passenger time valuation

In basic microeconomic models, the only time-allocation decision to be taken is to determine the amount of leisure. Leisure and market working time are complements which add up to the available time budget. Passenger travel time is then calculated as "leisure" and its value is directly and exclusively determined by the wage rate.

The aggregate evaluation of travel time savings for a population affected by a certain transport policy measure would then depend on the opportunity costs to the users of the transport system. These costs are the result of summing individual times spent travelling multiplied by the respective real incomes.

The most influential deviations from this standard approach (DeSerpa, 1971; MVA Consultants 1994, Mackie *et al.*, 2001) suggest that the different uses of time should possibly have a direct benefit, other than through the income effect, by varying the market working time. Passenger travel time would then depend not only on real incomes but also on the direct benefits of spending time in transport. Low valuations of travel time would therefore be interpreted as direct benefits of travelling, partly compensating for the opportunity costs of transport. This idea has recently met with renewed interest (Handy *et al.*, 2005; Ory and Mokhtarian, 2005).

What distinguishes the vast literature following on from DeSerpa is the fact that it is based on the assumption that consumers fail to achieve "technical efficiency in transport". In other words, the transport demand of an individual has a certain technically defined, minimum time requirement. What characterises the work following DeSerpa from the standard labour supply model is the assumption that individuals systematically spend more time in transport than is technically required.

At least two features of this approach appear to be dissatisfactory:

– The assumption of the individual's failure to optimise consumption decisions appears to be problematic from a methodological point of view: the fact that it is couched in a neoclassical framework, with perfect information and perfect ability to optimise, leads to the question: which characteristics of travel decisions are responsible for consumers' failure to optimise only when allocating time to transport decisions? Given the basic assumptions of full information, unconstrained information processing ability and rationality, defined in a purely substantive rather than procedural sense, the DeSerpa model is based on auxiliary assumptions (limits to optimisation in transport) which violate the basic assumptions of the theoretical framework[1].

The complexity of transport demand decisions might, of course, imply that substantive rationality is unachievable and that transaction costs of information collection and processing defy substantive rationality, but this might hold for other consumption choices as well, and should be modelled.

– If there are transaction costs leading to consumer decisions which indeed do not minimise travel time, the question of what this implies for the rationality of transport policy decisionmaking has to be answered. A failure to identify the minimum travel times would render the DeSerpa approach to identifying the value of time impractical. Information costs for planners to identify minimum travel times may be formidable. Shop scheduling and vehicle routing belong to the NP hard problems, which are impossible to solve algorithmically in finite time. Without an empirically measurable and measured difference between the actual and minimum transport times, the basic result obtained from the DeSerpa approach is identical to the conclusion of the basic microeconomic model, i.e. the value of travel time is identical to the real wage rate.

In planning, the idea that individuals fail to achieve minimum travel times has often been taken for granted. It has recently found renewed resonance in the literature (Handy *et al., 2005*), distinguishing between "driving by choice" and "driving by necessity" and raising the question: in what ways do people drive more than they really need to, thereby generating "excess driving"? "Excess driving" is defined as car use which goes beyond what is required for household maintenance, given choices about residential location, job location and activity participation. More specifically, the required level of driving is defined by the minimum number of trips, using the shortest routes to the closest destinations possible, and using other modes than the car as often as possible (*ibid.*, p. 185). Planners are supposed to aim at the reduction of "excess driving", by helping to reduce travel times for trips that are carried out from necessity rather than choice. To be operational, such a proposal requires that transport policy planners would indeed have superior information on the shortest trips of individuals, would completely understand the motivations of passengers, for example how multi-purpose trips are planned, and would need enormous information processing capacities. Otherwise, restricting the sovereignty of consumers' choices might entail substantial welfare losses.

Despite some reference to Becker's theory of the allocation of time (Becker, 1965), there exists only a small number of studies actually applying it to the identification of the value of travel time (cf. Gronau, 1971). In this stream of literature, the derivation of the value of time is cast under the category, "household production theory" or "theory of home production" (Gronau, 1977). According to this approach, consumers face time and budget constraints, and welfare is not a direct function of market goods and services but of "commodities". These "commodities" are "produced" by households by combining market goods and services (bought by spending wage income earned by allocating time to market work) with time. Market goods and services are, in turn, bought by spending wage income which has been earned by allotting time to market work.

The aggregation of leisure and work time at home, which would lead to the evaluation of travel time with the wage rate, is rejected, as it has to be based on the assumption that the two elements:

a) react similarly to changes in the socioeconomic environment and therefore nothing is gained by studying them separately, and

b) satisfy the conditions of a composite input, i.e. their relative price is constant and there is no interest in investigating the composition of the aggregate, since it has no bearing on household production and the shadow value of commodities.

Basically, all time-budget studies (for a recent example, see Hammermesh and Gronau, 2003) have established that time uses for consumption and work at home are not affected in the same way by socioeconomic variables. They have also shown that the composition of the aggregate affects many facets of consumption behaviour.

According to the study by Gronau and Hammermesh (2003), all commodities in the sense of household production theory are produced by goods and services as well as time, excepting sleep. That is, travel expenditure and travel time produce the commodity "travel". It is worth noting that the travel commodity is relatively goods-intensive -- i.e. household expenditure on transport is relatively high -- in contrast to the view that travel time alone and in isolation is a commodity, as assumed in the studies following DeSerpa's pioneering article. The goods intensity of travel and, with it, the shadow value of time differ systematically according to the socioeconomic characteristics of the consumer. Interestingly, and in contrast to other commodities, travel expenditure intensity does not increase with schooling and income. Once one moves beyond the lowest level of education, goods and time inputs into travel move in the same proportion as changes in the average goods and time inputs into all home-produced commodities. At relatively low income levels, the increase in the goods intensity of travel is due to the shift from public to private transport. The share of commodity travel in the overall time budget (travel time plus the market working time needed to earn the expenses for travel) moves negatively with the commodity "childcare" over the lifecycle and is otherwise constant. The travel variable excludes time and expenditure on commuting travel, which echoes the distinctions of "travel by choice" and "travel by necessity".

However, general concerns about the applicability of home production seem to be particularly relevant for empirical studies on the value of travel time (Pollak and Wachter, 1975; Pollak, 2002). The tractability of the models following Becker's original work on the allocation of time depend, however, crucially on two assumptions:

–   the absence of joint production;
–   the observability and measurability of commodities.

Recent research on the input or output characteristics of passenger transport (e.g. Ory and Mokhtarian, 2005) casts doubts on both auxiliary assumptions. Many of the reasons for travel given by passengers raise the question of how the commodity "passenger travel" and its attributes could be reliably measured. What seems to emanate from this literature is the fact that travel time leads to more than one commodity with the implication of households' "joint production". For example, getting from location A to location B and being able to work on the computer or listen to music implies joint output, which makes it difficult to interpret empirical findings. With joint production, commodity shadow prices -- i.e. the number of units of the *numéraire* the consumer is willing to give up to receive one more unit of the commodity -- depend not only on the resources available to the household and its opportunities to produce commodities but on consumer preferences as well. In other words, with joint production the passenger becomes a monopsonist with a non-linear budget constraint. Hence, the analogy with standard consumer theory disappears. Empirically, it becomes difficult to sort out whether commodity prices reflect differences in tastes or differences in opportunities.

As was evident from the Round Table discussions, the debate on how to assess the value of time, whether by observed passenger choices or by stated preferences, continues. The Round Table background papers and the discussion highlighted the fact that differences persist in the values of time obtained from revealed preference studies and stated preference studies.

In contingent valuation studies, respondents are usually given information on specific, hypothetical time delays. They are also confronted with a question or questions which provide information about the economic sacrifice they would have to make in order to reduce time losses. This may take the form of an open-ended question, asking what is the maximum amount they would be willing to pay for reductions in travel times. It might involve a series of questions confronting them with different prices for different speed increases, depending on their previous answers, or it may take the form of a hypothetical referendum in which respondents are informed of how much each would have to pay if an infrastructure charging regime to reduce travel times were implemented (Portney, 1994).

Compared to revealed preference studies and evidence from laboratory experiments, contingent valuation studies seem to systematically overstate values of travel time. Normally, however, there is no data on actual transactions to compare with survey responses indicating hypothetical willingness to pay. Concerns about contingent valuation include a credibility bias (or "reliability" of responses) and the level of precision of the answers. Credibility refers to whether survey respondents are actually answering the question posed, and reliability refers to the size and direction of the biases that may be present in the responses. Precision relates to the variability of the answers, a problem which can be dealt with by increasing the sample size. Therefore, the precision problem is one which concerns the resources to be made available to studies in order to render the results viable.

A first concern about the contingent valuation results is the fact that they are at variance with rational choice. While one might wonder why such "embedding effects" (Kahneman and Knetsch, 1992) are relevant for contingent valuation methods, rationality requirements (concavity of preferences) seem necessary to ensure that the verbal answers have some degree of consistency. The fact that willingness-to-pay for time savings is overstated may have to do with the fact that other, competing resource demands are excluded from the respondent's considerations. More generally, respondents seem to fail to think carefully about how much disposable income they have available to allocate to all ends.

If the valuation of passenger travel time is included in stating preferences for infrastructure projects with the implication of hypothetical time savings, respondents may fail to fully register information given on the infrastructure project and its associated time savings. Respondents often tend to differ with respect to estimates of the consequent time savings. Finally, they sometimes answer by signalling their rating of the general importance of a certain project to society. Some respondents appear to answer with zeros when asked to reveal their support for time-saving transport policies as a way of conveying their objections. This has led critics of contingent valuation methods to conclude that answering survey questions may fulfil the same function as contributing to charitable institutions -- not only to support a certain project or organisation but to feel the "warm glow" which comes from supporting worthy causes (Arrow *et al.,* 1993; Diamond and Hausman, 1994; and Andreoni, 1989).

Closely related to the assessment of the value of time is the discussion of travel time budgets and budgets for monetary expenditure. The idea of constant travel times per day (or per week) has gained much influence on transport planning. It has become so popular and suggestive that the word "budget" is held to imply stability, referring to an *"allocation of time, money or generalised resources to travel which would not be influenced by policy, trends or costs"* (Goodwin, 1981, p. 97; Mokhtarian and Chen, 2004).

The research into the constancy of travel times emanated from dissatisfaction with standard (four-step) models to forecast regional travel demand. This dissatisfaction was due to the inability of the standard forecasting models to take behavioural changes properly into account, leading to difficulties in fitting the model to observed data. While the hypothesis of stable travel times had been around for some years, it became widely known through Zahavi's "Unified Mechanism of Travel" (UMOT) process for forecasting passenger transport (Zahavi, 1979). The UMOT concept started out from the expectation that travel times (and expenditures) would be robustly associated with household characteristics and supply conditions of the transport system. Zahavi's objective was to simplify the forecasting of passenger transport by avoiding "that a lengthy calibration process to observed data is required" (Zahavi and Talvitie, 1980b). The idea of constant travel time budgets won such a strong acceptance that a dispute emerged about its numerical value: 1.1–1.3 hours per day as the universal constant was maintained by Zahavi (Zahavi and Ryan, 1980a; Zahavi and Talvitie, 1980b; Hupkes, 1982); 1.1 hours per day was also supported by Bieber *et al.* (1994) and Schafer and Victor (2000), while Vilhelmson (1999) claimed that it is rather 1.3 hours per day. Hupkes (1982) confirmed the hypothesis that people have an unobserved, desired travel time budget per day and per person, which might be defined by a bio-physiological desire to maintain a fixed daily routine, with the implication of a fixed travel time budget of 430 hours of travel per person per year. Chumak and Braaksma (1981) argued that the concept of a constant travel time budget can be used to check conventional forecasting results, to ensure that those results reflect an equilibrium between travel demand and the supply of transport facilities. There was even the proposal to build constant travel time budgets into the components of conventional forecasting models (Goodwin, 1981). Only recently has the assumption of constant travel times been used to predict worldwide mobility, as incomes rise and slower modes are replaced by faster ones (Schafer and Victor, 2000).

The strong belief in the constancy of travel times led to the claim that transport policy actions were at best futile, if not dangerous. Transport policy measures, according to the adherents to the constant travel time hypothesis, would allow for higher speeds and increase the distances travelled. If based on models which associate time directly with utility, it is concluded that no benefits result, only an increase in environmental costs. Hence, sceptics warned that good intentions to reduce costs would ultimately be counterproductive, leading to social net losses (e.g. Noland and Lem, 2002). Rather, a large number of factors have been found to influence travel times (and therefore the value of time), with substantial differences according to the environments in which the studies have been carried out (Mokhtarian and Chen, 2004). However, Mokhtarian and Chen, who undertook the formidable task of comparing 22 studies on the constancy of passenger travel times per day, found that the hypothesis of its stability is not supported.

The results of the large number of studies which have been carried out on travel time budgets suggest that there is a "law" regarding their constancy per day and per person. Hence, the values for daily hours used for transport, which have been popularised in aggregate studies, should not be taken for granted. They do not obviate the need for careful studies on the value of time of passengers who are affected by transport policy measures.

3. TIME AND FREIGHT TRANSPORT

While values of time for passenger transport are routinely used in cost-benefit analyses (CBA) of transport policies, this is far less the case for time savings in freight transport. One reason for the difference in the valuation of passenger and freight time conventions may be seen in the lack of an obvious analogy between the theoretical background of household production and passenger values of time for freight transport. The value of freight transport time, as reflected in the capital costs of the goods in transit, should be fully reflected in the market price of the final consumer good. The CBA of transport policies leading to shorter goods transport times would capture the value added through the shorter transport times by calculating the increase in consumers' rents arising from a reduction in the price of the respective consumer good.

The Round Table discussion showed that the rationale for valuing freight transport derives from the often neglected effects of the speed of transport on logistics costs, on the relative competitiveness of industrial production locations and on the relocation of firms. An attempt can be made to capture the effects of freight transport improvements by defining "generalised freight transport costs", which include the time costs of freight transport. The generalised costs of transport -- in addition to the wage costs of the driver and the energy and vehicle costs -- include three cost components for valuing logistics improvements:

- the time costs of goods in transit;
- the cost of uncertainty about the duration of the freight transport time;
- the expected costs of damage to goods in transit.

The conceptual discussion focuses on how to define the unit cost values for these cost categories: the value of one time unit of delay of a quantity unit of freight; the value of cost reductions due to a smaller size buffer stock to smooth the volatility of transport times; and the value of the expected loss of goods in transit.

Even on the conceptual level, most of the studies have concentrated on defining unit values for delays in transport. The approaches to defining unit values can be further divided into a "capital cost" approach and a "stated and revealed preference" approach.

3.1. Capital cost approach

The capital cost approach relies on three parameters in order to define the value of time, i.e. the rate of interest; the value of goods being carried per tonne and number of tonnes being transported per vehicle; and the number of hours in transit per year. The most frequent explanation for using the capital cost evaluation when defining unit values for freight transport time is that reduced transport times lead to reduced working capital costs. These costs are then considered to reflect reduced resource costs resulting from higher speeds in freight transport.

Critics of this approach maintain that it underestimates the value of time. In fact, a higher willingness to pay for a more rapid delivery of goods may not always induce greater speed of delivery, as supply responses face institutional or infrastructure-related constraints.

The value of the improved reliability of freight transport has only been discussed on a conceptual level (Minken, 1997). So far this conceptual work has not been used for empirical estimates of the value of expected transport time savings due to improved freight transport reliability. The approach proceeds from a description of the variability of transport times in terms of a probability density function and its impact on optimal inventory planning. The value of increased reliability is derived from the cost reductions associated with smaller buffer stocks, in order to optimally respond to the variability of transport times. These values are computed by using a simulation model, which reflects the relationship between transport time variability and buffer stock costs. In an early study (Bruzelius, 1986), it was estimated that fully taking account of an improved reliability of freight transport would lead to a doubling of conventional estimates of the value of freight time, based on the capital cost approach.

Even less progress has been made on including values of reduced damage as a component of the value of freight time into transport policy evaluation studies. Individual studies have identified very low values of expected damage due to shorter transport times.

3.2. Stated and revealed preference analyses

The vast majority of studies on the value of time have been Stated Preference and Revealed Preference studies. The background report to the Round Table (Tavasszy, 2005) mentions 36 studies empirically estimating the value of time. The studies conclude with values of time which are far higher than the values obtained from capital value studies.

The most common approach to estimating the value of reliability is by Stated Preference Studies. Reliability is measured by the percentage delay of the average transport time. Other definitions of reliability include measurement by the spread between the earliest arrival time of a given shipment and the time when 98 per cent of the shipment has arrived (Fowkes *et al.*, 2001).

The elaborate approach of Small *et al.* (2005) to identify commuters' preferences for reliable highway travel has so far not been applied to freight transport. It is based on a simulation model (Noland *et al.*, 1998) to determine the impact on congestion of policies for dealing with travel time uncertainty. In combining revealed and stated preference data, the model combines a supply-side model of congestion delay with a discrete choice econometric demand model, which predicts choices for trips with an ideal arrival time. The strong current interest in identifying the benefits of a greater reliability of travel times has led to the expectation of substantial progress in evaluating the related benefits of transport policy in the near future.

Implementation of unit time values for freight transport

In contrast to the widespread application of valuation standards used in cost-benefit analyses of transport policy measures, freight transport valuations are nowhere applied, with one exception[2]. Only the transport authorities of Sweden -- the National Roads Authority (VV) and the National Rail Authority (BV) -- have integrated goods-specific unit values for freight transport into their evaluation methodologies. These unit prices have also been used for their SAMGODS forecasting model by the Swedish SIKA Institute.

In other countries, work on including freight time valuation standards in national methodologies of cost-benefit analyses has not gone beyond the research stage. In the Netherlands and the United Kingdom, major studies on the valuation of freight time were carried out in the 1990s without having much influence on planning and evaluation in practice.

At the European level, a project commissioned under the Fourth Framework Programme, EUNET (1998), had the objective of defining a methodology for evaluating transport infrastructure investment. The project contained a section on the valuation of freight transport time. A report of the Western European Road Directorates (2000) recommends the use of the EUNET freight values of time.

First steps have also been taken in Australia and New Zealand to implement vehicle-specific freight transport time values (Austroroads, 2000).

Interestingly, the Swedish approach to taking account of freight transport values of time is based on the capital value approach. In view of the Round Table discussion on evaluation standards, this is likely to imply an under-valuation of freight transport time and, consequently, an under-evaluation of freight transport time savings achieved by transport policy actions.

Secondary effects of freight transport time savings and value of time

Beyond the induced organisational changes in the logistics sector, the time requirements for freight transport in the wider economic context have only recently attracted the attention of researchers and policymakers: time savings in freight transport which follow from transport policy action have the potential to substantially alter the interregional division of labour and, in the longer term, to induce new patterns of industrial location.

These additional time-of-freight-transport effects, which go beyond the immediate logistics effects, concern the limits of the partial equilibrium approach of standard CBA. A reduction in transport times and the associated reduction in transport costs might lead to the strengthening of regional specialisation on economic activities in which they have comparative advantages in production. Moreover, the increased trade between regions can aid the diffusion of technical and organisational knowledge, which in turn increases productivity in the trading regions.

In the case of the prevalence of industries where larger firms have a cost advantage over smaller firms, the reduction in transport costs will strengthen the agglomeration economies of large economic regions. In a number of countries, the United Kingdom and the Netherlands in particular, transport planning has taken the first steps towards taking account of the income effects of transport time savings on the interregional division of labour and the relocation of manufacturing industries. The

Federal Highway Administration's "Freight Benefit Cost" White Paper (FHWA, 2001) gives an account of the microeconomics of organisational responses to freight transport time savings and a decrease in transport costs more generally.

The development of comprehensive planning models to deal with the secondary effects is still in its infancy. The spatial computable equilibrium models which have been developed to evaluate transport policies in general and transport infrastructure investment in particular (e.g. IASON, 2002), use rather restrictive assumptions on manufacturing production and the transport sector[3].

It is to be expected that the value of freight time transport which takes into account secondary effects will be considerably greater than that derived from a partial equilibrium cost-benefit analysis. Whether unit values for freight transport time will be systematically fed back into routine cost-benefit analyses of transport policies will depend on the future development of multi-sectoral and multi-regional models, the availability of the required data and the quality of the parameter estimates, as well as the acceptance of the advanced planning tools in the policy decisionmaking process. An important aspect of the increased importance of secondary effects has to do with globalisation. The costs and time requirements of international transport play an important role in determining the pattern of bilateral trade flows.

4. VALUE OF FREIGHT TRANSPORT TIME AND INTERNATIONAL TRADE

A specific indication of the significance of the secondary effects of a more efficient transport sector can be seen in the importance that trade costs, and transport costs as part of the trade costs, have gained for the analysis of international economic relations. Traditionally, the theory of international trade has paid little attention to transport costs. A main reason for this neglect was the incompatibility between the standard view that international goods markets are perfectly competitive and the existence of substantial trade costs. To the extent that transport costs have been taken into account, a special type has been assumed: to avoid the modelling of transport as a separate economic sector, the resource costs of transport have been assumed to accrue as a percentage of the good shipped.

In view of the reduction of tariffs, and to a lesser extent of non-tariff barriers to trade, researchers studying international economic relations were not so much surprised by their rapid expansion, or "globalisation", as by the fact that it did not go further. Part of this puzzle could be solved by taking a closer look at technological differences between trading partners (Trefler, 1993). But even taking account of the differences in technology, actual goods trade represents only a small fraction of what theory would predict as the volume of frictionless trade (Trefler, 1995). One possibility for explaining the "Mystery of Missing Trade" could lie in the existence of substantial trade costs.

An indication of the importance of transport costs as part of the trade costs is provided by recent estimates of gravity equations for international trade. A gravity equation relates the volume of trade between pairs of countries positively to their economies' sizes and inversely to the distance between them. Estimates of gravity equations consistently find an elasticity of trade with respect to distance of around 1.0. This means that countries which are twice as far apart trade half as much on average.

Time cost is only one element of transport costs in international trade. Insurance costs, the costs of trade financing and costs of legal advice might all be considerably higher when borders are crossed than in the case of national transport. Some of the resource costs will also be time-dependent. A first time-dependent component is the interest cost which derives from the fact that the supplier incurs the cost of production before a good is shipped, but is paid for the shipment only after the delivery. The cost of this time delay is the interest paid or foregone by the shipper in order to pay the workers and suppliers before being paid by the customer. Storage and depreciation costs will add to the interest costs. They will be higher the greater the uncertainty of demand.

The main importance of time requirements in international transport does, however, follow from the increasing speed of technical change and the shortening of fashion cycles. As long as technical change is slow and consumer preferences remain unchanged, long transport delays only increase storage and depreciation costs, which usually have a small share in overall value added. High-technology products with short product cycles (Vernon, 1966) and goods with short periods of consumer demand due to changing fashions (Evans and Harrigan 2005), might become entirely obsolete in transit. Evans and Harrigan (2005) show that, for fashion goods, production has been moved from more distant locations with lower production costs to closer locations, even if production costs are higher.

The decreasing time costs for international transport, following a massive switch from maritime to air transport, seems to have been more important as a driver of globalization than decreasing resource costs (Hummels, 2001). In contrast to widespread belief, the resource costs of transport have fallen far less than would be required to explain the international trade expansion of recent years (Hummels, 1999).

The Round Table discussion emphasized that neglect of the particular characteristics of time costs in international transport might cause high losses of potential income gains from international trade. The income losses due to transport costs are even more severe than their description in terms of ad valorem tariffs might indicate: tariff revenues which reflect a transfer from the consumer to governments compensate to some extent the income losses resulting from market friction. They have no equivalent in the case of high transport costs.

The fact that little attention had been paid to the importance of transport for international trade suggests that the benefits of transport policies to reduce time requirements for international freight transport have been substantially underestimated. In view of the Round Table discussion, the inclusion of valuation standards for international freight transport time is even more important than for national domestic transport.

5. CONCLUSION

The evaluation of time savings in passenger and freight transport is often crucial for the acceptance or rejection of transport policies or projects. Despite the abundant literature that has developed on the evaluation of passenger travel time, no consensus on valuation standards has emerged. The value of time savings in freight transport is rarely taken into account in transport planning.

In current studies on the evaluation of passenger travel times, it is surprising to observe the low value which consumers of transport services seem to assign to time in transport. Moreover, there are substantial differences in values of time in transport derived from observed behaviour (revealed preferences) and responses to questionnaires (stated preferences). More reliable estimates of values of passenger time depend, firstly, on more comprehensive conceptual foundations of estimation models in the tradition of the revealed preference literature. Secondly, at least in cases where no past events comparable to the evaluated transport policies exist, stated preference studies will remain important. They should make full use of the progress which has recently been achieved in contingent valuation methods.

The presumption of a fixed time budget has led to concerns that time savings achieved by transport policy actions are partly undone by the reactions of passengers, with negative consequences for the environment. The Round Table assessed the evidence indicating that the hypothesis of constant travel time budgets is difficult to maintain. Travel time evaluations should not proceed from an assumption of constancy of time budget. Forecasts of travel demand based on this assumption tend to underestimate the net benefits of transport policies.

Time savings in freight transport are largely neglected in transport planning. Although there are a sizeable number of studies on the cost and benefits of freight transport time savings, the adoption of standards for evaluating transport policy remains an exception. Where such standards have been adopted, as in Sweden, they are based on a narrow approach, the capital cost approach. Within this concept, freight time costs are defined as working capital costs and depreciation in transit. Revealed and stated preference studies on freight transport time showed, in contrast, far higher values.

The Round Table discussion emphasized the importance of taking into account the induced changes in supply chain organisation. For major policy interventions and transport infrastructure projects, changes of freight transport speed can have a strong impact on the interregional division of labour and the location decisions of manufacturing industries. Routine cost-benefit analyses tend to neglect the substantial income effects which result from such secondary consequences of transport policies.

A particular aspect of these secondary effects is the consequences of international transport speed at times of globalisation. The empirical evidence shows that it was the speed of international transport, rather than savings in resource costs, which has worked as a determinant for the intensification of international economic relations. Given the shortening of product and fashion cycles, time requirements for international transport play an increasing role in determining the geographic pattern of transport demand. To manage these developments requires sustained efforts in international transport policy co-ordination.

NOTES

1.   On the methodological discussion on the necessity and realism of assuming the substantive rationality of consumers, cf. Caldwell (1994).

2.   This conclusion, from the survey of Walters *et al.* (1995), still holds today.

3.   See, for example, the analysis of the links between the efficiency of the transport sector and the general economic development, in Lakshmanan and Anderson (2002), indicating the limitations of current Computable General Equilibrium Models.

REFERENCES

Andreoni, J. (1989), Giving with impure altruism: Applications to charity and Ricardian equivalence, *Journal of Political Economy*, 97: 1447-1458.

Arrow, K., R. Solow, P.R. Portney, E. Leamer, R. Radner and H. Schuman (1993), Report of the NOAA panel on contingent valuation, *Federal Register*, 58: 4602-4614.

Becker, G.S. (1965), A theory of the allocation of time, *Economic Journal*, 75: 493-517.

Caldwell, B.J. (1994), *Beyond Positivism: Economic Methodology in the Twentieth Century*, Oxford.

Chumak, A. and J.P. Braaksma (1981), Implications of the travel-time budget for urban transportation modeling in Canada, *Transportation Research Record* (794): 19-27.

Crozet, Y. (2005), Time and passenger transport, *Round Table 127: Time and Transport*, OECD/ECMT, Paris.

Deardorff, A.V. (2005), The importance of the cost and time of transport for international trade, *Round Table 127: Time and Transport*, OECD/ECMT, Paris.

DeSerpa, A. (1971), A theory of the economics of time, *Economic Journal*, 81: 828-845.

Diamond, P.A. and J.A. Hausman (1994), Contingent valuation: Is some number better than no number?, *Journal of Economic Perspectives*, 8: 45-64.

Evans, C. and J. Harrigan (2005), Distance, time, and specialization: lean retailing in general equilibrium, *American Economic Review*, 95: 292-313.

Goodwin, P. (1981), The usefulness of travel budgets, *Transportation Research A*, 15: 97-106.

Gronau, R. (1971), The effect of travelling time on the demand for passenger transportation, *Journal of Political Economy*, 79: 377-394.

Gronau, R. (1977), Leisure, home production, and work -- the theory of the allocation of time revisited, *Journal of Political Economy*, 85: 1099-1123.

Gronau, R. and D.S. Hamermesh (2003), Time versus goods: the value of measuring household production technologies, NBER Working Paper No. 9650, Cambridge, Mass.

Handy, S., L. Weston and P.L. Mokhtarian (2005), Driving by choice or by necessity?, *Transportation Research A*, 39: 183-203.

Hummels, D. (1999), Have international transportation costs declined?, Purdue University.

Hummels, D. (2001), Time as a trade barrier, mimeo, Purdue University, Lafayette.

Hupkes, G. (1982), The law of constant travel time and trip-rates, *Futures*: 38-46.

IASON (2002), Methodology for the assessment of spatial economic impacts of transport projects and policies, Delft.

Kahneman, D. and J. Knetsch (1992), Valuing public goods: The purchase of moral satisfaction, *Journal of Environmental Economics and Management*, 22: 57-70.

Lakshmanan, T.R. and W.R. Anderson (2002), Transportation Infrastructure, Freight Services Sector and Economic Growth. White Paper prepared for the US DOT/FHWA. Center for Transportation Studies, Boston.

Mackie, P.J., S. Jara-Diaz and A.S. Fowkes (2001), The value of travel time savings in evaluation, *Transportation Research, Part E*, 37: 91-100.

Mokhtarian, P.L. (2005), Travel as a desired end, not just a means, *Transportation Research, Part A*, 39: 93-96.

Mokhtarian, P.L. and C. Chen (2004), TTB or not TTB, that is the question: a review and analysis of the empirical literature on travel time (and money) budgets, *Transport Research, Part A*, 38: 643-675.

MVA Consultancy, Institute for Transportation Studies at Leeds University, and Transport Studies Unit, Oxford (1994), Time savings: Research into the value of time, in: R. Layard and S. Glaister (eds.), *Cost-Benefit Analysis*, 2nd rev. ed., Cambridge, Mass.

Noland, R.B. and L.L. Lem (2002), A review of the evidence for induced travel and changes in transportation and environmental policy in the US and the UK, *Transportation Research D*, 7: 1-26.

Ory, D.T. and P.L. Mokhtarian (2005), When is getting there half the fun? Modeling the liking for travel, *Transportation Research A*, 39: 97-123.

Pollak, R.A. (2002), Gary Becker's contributions to family and household economics, NBER Working Paper No. 9232, Cambridge, Mass.

Pollak, R.A. and M.L. Wachter (1975), The relevance of the household production function and its implications for the allocation of time, *Journal of Political Economy*, 83: 255-277.

Portney, P.R. (1994), The contingent valuation debate: Why economists should care, *Journal of Economic Perspectives*, 8: 3-17.

Schafer, A. and D.G. Victor (2000), The future mobility of the world population, *Transportation Research A*, 34: 171-205.

Tavasszy, L. (2005), Freight and Transport, *Round Table 127: Time and Transport*, OECD/ECMT, Paris.

Trefler, D. (1993), International factor price differences: Leontief was right, *Journal of Political Economy*, 101: 961-987.

Trefler, D. (1995), The case of the missing trade and other mysteries, *American Economic Review* (1029-1046).

Vernon, R. (1966), International investment and international trade in the product cycle, *Quarterly Journal of Economics*, 111 (190-207).

Vilhelmson, B. (1999), Daily mobility and the use of time for different activities: the case of Sweden, *GeoJournal*, 48: 177-185.

Waters, W.G., C. Wong and K. Megale (1995), The value of commercial vehicle time savings for the evaluation of highway investments: a resource saving approach, *Journal of Transportation Research Forum*, 35.

Zahavi, Y. (1979), UMOT Project. Prepared for US Department of Transportation, Washington DC and Ministry of Transport, Federal Republic of Germany, Bonn. Report DOT-RSPA-DPB-20-79-3.

Zahavi, Y. and J.M. Ryan (1980), Stability of travel components over time, *Transportation Research* (750): 19-27.

Zahavi, Y. and A. Talvitie (1980), Regularities in travel time and money expenditures, *Transportation Research Record* (750): 13-19.

LIST OF PARTICIPANTS

Prof. Eddy VAN DE VOORDE   **Chairman**
University of Antwerp
Faculty of Economics
Department of Transport and Regional Economics
Prinsstraat 13
B-2000 ANTWERP 1
Belgium

Prof. Yves CROZET   **Rapporteur**
Directeur
Laboratoire d'Economie des Transports (LET)
Université Lumière Lyon 2
MRASH
14 avenue Berthelot
F-69363 LYON Cedex 07
France

Prof. Alan DEARDORFF   **Rapporteur**
The University of Michigan
Gerald R. Ford School of Public Policy
Department of Economics
440 Lorch Hall
611 Tappan Street
ANN ARBOR, Michigan 48109-1220
United States

Dr. L.A. TAVASSZY   **Co-rapporteur**
TNO INRO
Schoemakerstraat, 97
P.O. Box 6041
NL-2600 JA DELFT
The Netherlands

Dr. Nils BRUZELIUS   **Co-rapporteur**
Nils Bruzelius AB
Mätaregränden 6
S-226 47 LUND
Sweden

Professor Michael BROWNE
University of Westminster
Transport Studies Group
35 Marylebone Road
GB-LONDON, NW1 5LS
United Kingdom

Mr Jorgen CHRISTENSEN
Director
Danish Road Institute (SVL)
Elisagaardsvej 5
DK-4000 ROSKILDE
Denmark

M. Jacques DELSALLE
Administrateur
CE/EC
DG Affaires Economiques et Financières
BU1 2/130
rue de la Loi 200
B-1049 BRUXELLES
Belgium

Mr. Jan FRANCKE
Ministry of Transport, Public Works and Water Management
Transport Research Centre (AVV)
P O Box 1031
NL-3000 BA ROTTERDAM
The Netherlands

Prof. Phil GOODWIN
Director, TSU
University College London
Centre for Transport Studies
22 Gower Street
GB-LONDON, WC1E 6BT
United Kingdom

Mr. Hugh GUNN
RAND Europe - Leiden
HGA
Weipoortseweg 69
NL-2381 NG ZOETERWOUDE
The Netherlands

Professeur Vincent KAUFMANN
Ecole Polytechnique Fédérale de Lausanne
Laboratoire de Sociologie Urbaine LaSUR
Bâtiment Polyvalent
CH-1015 LAUSANNE
Switzerland

Prof. Peter KLAUS
Friedrich-Alexander-Universität
Lehrstuhl für BWL, insbes. Logistik
Theodorstrasse 1
D-90489 NÜRNBERG
Germany

Mme Evdokia MOÏSÉ
Trade Directorate
OECD
2 rue André Pascal
F-75775 PARIS CEDEX 16
France

Prof. Patricia MOKHTARIAN
University of California at Davis
Dept of Civil and Environmental Engineering
1 Shields Avenue
DAVIS, CA 95616
United States

Mr. Janos MONIGL
Managing Director
Transman Consulting Ltd
Hercegprimas u.10
H-1051 BUDAPEST
Hungary

Monsieur André PENY
Responsable Mission Transport
Ministère de l'Equipement, des Transports et du Logement
DRAST
Tour Pascal B - Paroi Sud
F-92055 LA DEFENSE CEDEX 04
France

Prof. Dr. Karin PESCHEL
University of Kiel
Institute for Regional Research
Olshausenstr. 40
D-24098 KIEL
Germany

Prof. Marco PONTI
President
TRT Trasporti e Territorio SRL
Via Rutila, 10/8
I-20146 MILANO
Italy

M. le Professeur Emile QUINET
Chef du Département
Ecole Nationale des Ponts et Chaussées
Département d'Economie et des Sciences
28 rue des Saints-Pères
F-75007 PARIS
France

Mr Markus RADL
Adviser
Federal Ministry for Transport, Innovation and Technology
Department K6 / EU-Affairs
Radetzkystrasse 2
A-1031 WIEN
Austria

Ms Charlene ROHR
RAND Europe - Leiden
RAND Europe
Grafton House
64 Maids Causeway
GB-CAMBRIDGE CB5 8DD
United Kingdom

Prof. Dr. Wlodzimierz RYDZKOWSKI
Chairman of Department
University of Gdansk
Department of Transportation Policy
Armii Krajowej 119/121
PL-81-824 SOPOT
Poland

Mme Cecile SEGONNE
Economiste
SNCF
Direction de la Strategie
34 rue du Commandant Mouchotte
F-75699 PARIS Cedex 14
France

Mr Juha TERVONEN
Consultant to the MOTC Research Unit
JT-Con
Hameentie 16 A18
FIN-00530 HELSINKI
Finland

Mr Mateu TURRO
Associate Director
Banque Européenne d'Investissement
100 Bld Konrad Adenauer
L-2950 LUXEMBOURG
Luxembourg

Prof. José Manuel VASSALLO
Escuela Tecnica Superior de Ingenieros de Caminos
Ciudad Universitaria, s/n
E-28040 MADRID
Spain

Dr. Mark WARDMAN
University of Leeds
Institute for Transport Studies
36 University Road
GB-LEEDS, LS2 9JT
United Kingdom

Dr. Staffan WIDLERT
Director General
SIKA (Statens institut för kommunikationsanalys)
Box 17213
SE-10462 STOCKHOLM
Sweden

Mr. Tadashi YASUI
Trade Directorate
OECD
2 rue André Pascal
F-75775 PARIS CEDEX 16
France

ECMT SECRETARIAT

Mr. Jack SHORT, Secretary General

ECONOMIC RESEARCH AND STATISTICS UNIT

Dr. Andreas KOPP, Head of Unit
Dr. Michel VIOLLAND, Administrator
Mrs. Julie PAILLIEZ, Assistant
Mlle Françoise ROULLET, Assistant

TRANSPORT POLICY UNIT

Mr. Masatoshi MIYAKE, Consultant

ALSO AVAILABLE

Transport and Economic Development. Series ECMT – Round Table 119 (2002)
(75 2002 10 1 P) ISBN 92-821-1298-5

What Role for the Railways in Eastern Europe? Series ECMT – Round Table 120 (2002)
(75 2002 04 1 P) ISBN 92-821-1371-X

Managing Commuters' Behaviour - a New Role for Companies. Series ECMT – Round Table 121
(2002)
(75 2002 11 1 P) ISBN 92-821-1299-3

Transport and exceptional public events. Series ECMT – Round Table 122 (2003)
(75 2003 04 1 P) ISBN 92-821-1305-1

Vandalism, Terrorism and Security in Urban Passenger Transport. Series ECMT – Round Table 123 (2003)
(75 2003 07 1 P) ISBN 92-821-0301-3

Transport and Spatial Policies: The Role of Regulatory and Fiscal Incentives. Series ECMT – Round Table 124 (2004)
(75 2004 09 1 P) ISBN 92-821-2321-9

European Integration of Rail Freight Transport. Series ECMT – Round Table 125 (2004)
(75 2004 06 1 P) ISBN 92-821-1319-1

Airports as Multimodal Interchange Nodes. Series ECMT – Round Table 126 (2005)
(75 2005 03 1 P) ISBN 92-821-0339-0

16th International Symposium on Theory and Practice in Transport Economics. 50 Years of Transport Research (2005)
(75 2005 05 1 P) ISBN 92-821-2333-2

To register for information by email about new OECD publications: www.oecd.org/OECDdirect
For orders on line: www.oecd.org/bookshop
For further information about ECMT: www.cemt.org

OECD PUBLICATIONS, 2, rue André-Pascal, 75775 PARIS CEDEX 16
PRINTED IN FRANCE
(75 2006 01 1 P) ISBN 92-821-2329-4 – No. 54519 2005